WITHDRAWN FROM
KENT STATE UNIVERSITY LIBRARIES

COMPUTER ARITHMETIC

WILEY SERIES IN COMPUTING

Consulting Editor

Professor D.W. Barron
Computer Studies Group, University of Southampton, UK

BEZIER · Numerical Control - Mathematics and Applications
DAVIES AND BARBER · Communication Networks for Computers
BROWN · Macro Processors and Techniques for Portable Software
PAGAN · A Practical Guide to Algol 68
BIRD · Programs and Machines
OLLE · The Codasyl Approach to Data Base Management
DAVIES, BARBER, PRICE, and SOLOMONIDES · Computer Networks and their Protocols
KRONSJO · Algorithms: Their Complexity and Efficiency
RUS · Data Structures and Operating Systems
BROWN · Writing Interactive Compilers and Interpreters
HUTT · The Design of a Relational Data Base Management System
O'DONOVAN · GPSS - Simulation Made Simple
LONGBOTTOM · Computer System Reliability
AUMIAUX · The Use of Microprocessors
ATKINSON · Pascal Programming
KUPKA and WILSING · Conversational Languages
SCHMIDT · GPSS - Fortran
SPANIOL · Computer Arithmetic

COMPUTER ARITHMETIC
Logic and Design

Otto Spaniol
Institut für Informatik, Universität Bonn

JOHN WILEY & SONS
Chichester · New York · Brisbane · Toronto

This edition is published by permission of Verlag B G Teubner, Stuttgart, and is the sole authorized English translation of the original German edition.

Copyright © 1981 by John Wiley & Sons Ltd.

All rights reserved.

No part of this book may be reproduced by any means, nor transmitted, nor translated into a machine language without the written permission of the publisher.

British Library Cataloguing in Publication Data:

Spaniol, Otto
   Computer arithmetic. - (Wiley series in computing)
    1. Electronic digital computer
    2. Logic design
   I. Title
   621.3819'582   TK7888.3   80-41867

ISBN 0 471 27926 9

Printed in Great Britain

PREFACE

This book arose from a number of papers and seminars at Saarbrücken University. It contains a comparative description of quick and economical algorithms for performing arithmetic operations in digital computers.

Although practically all information scientists and not a few scientists in other branches are confronted in varying degrees with these problems, text books in the German language with a more general purpose deal with these problems only marginally or in relatively short chapters. This necessarily results in restriction to a description of the most familiar and simplest methods. There are several excellent standard works in the English literature (principally Flores "The logic of computer arithmetic"), but these are often difficult to get hold of and are usually out-dated. Hence several important results can be sought after in vain.

The lack of more recent books on computer arithmetics cannot be attributed to the fact that all the problems involved have already been treated in standard works. The number and relevance of works published most recently point to the opposite. The main cause in my opinion is rather the frequently heterogeneous representation and the close concentration of several works on a certain machine concept or on a range of components available at the time or being developed, so that the comprehensive treatment is rendered difficult.

Two main purposes of the present text are therefore a standard description as far as possible and, building on this, a comparative discussion of differing concepts which can be used when designing computers. The description of the interplay of the arithmetic unit with the other components of a computer fall into the background due to the scope, but these problems are treated comprehensively in other books dealing with computer architecture.

In addition to the familiar methods of performing arithmetic operations and special functions (trigonometric and hyperbolic functions, logarithms, roots etc.) the book investigates several matters for parallel operations which are still little used. Pipelining principles for accelerating and increasing the efficiency of arithmetic units are also discussed in detail. Two chapters deal with arithmetic in the case of redundant notations and complexity problems increasingly coming to the fore in recent times.

The design of algorithms is demonstrated by micro-programs and operational charts. The sequence and differentiation of the individual procedures are demonstrated on numerous examples.

The text is constructed in the manner of mathematics lectures. This deviates from the type of presentation usually chosen when treating this subject, but in my opinion both main purposes "uniformity" and "comparative treatment" are best achieved in this manner. Special knowledge beforehand is not essential for understanding the text.

I wish to thank my former colleagues Heinz Fuchs, Helmut Jäger and Hartmut Roos who have helped me in the preparation of the manuscript in addition to Miss Maria Vogelgesang for programming the examples to the CORDIC procedure. Thanks are due also to the Teubner-Verlag for their close co-operation and to Wiley and Co. for preparation of the English version and the excellent typing of the text.

Saarbrücken, May 1976
Bonn, September 1980   Otto Spaniol

C O N T E N T S

1.	Notation	10
1.1	denary digit codings	10
1.2	Insertion of longer notations; overflow problem	12
1.3	Arithmetic for denary digit systems	14
1.4	Other notations	19
1.5	Basic choice, register, connections, microprograms	20
1.6	Representation of fixed and floating points	22
2.	Adders	33
2.1	(m,k)-counter, halfadder, fulladder	33
2.2	Description of simple adder logic	36
	(notation, serial addition, von Neumann adder, carry save addition, adder-tree, carry ripple addition, asynchronous C-R-adder, exclusive-OR-adder)	
2.3	Carry look ahead addition	49
2.4	Carry skip addition	56
	(constant group size g, variable group size, higher order carry skip addition)	
2.5	Conditional sum addition	67
2.6	Carry select addition	72
2.7	Conclusion, comparison	75
3.	Multiplication	77
3.1	Register configuration, notation, overflow problems	77
3.2	Serial multiplication without multiplier coding	80
3.3	Multiplier coding	88

3.4	Unclocked and parallel methods of multiplication (multiplication matrix, reduction, multiplication by carry save addition, parallel multiplication according to Wallace and Dadda, investigation of effort, factors of various lengths, multiplication with base 2^h)	100
3.5	Arithmetic circuits	120
3.6	Pipelining principles	131
4.	Division	140
4.1	Fundamentals	140
4.2	Serial division methods for non-negative, binary operands (restoring, non-performing, non-restoring division, division with shifts over zeros and ones)	142
4.3	Negative operands	151
4.4	Acceleration of division by using suitable multiples of the divisor (table look up division, application of special divisor multiples)	155
4.5	Iterative division	173
5.	Redundant notation	194
5.1	SDNR with base $d \geq 3$	194
5.2	Parallel addition of SDNR addends	196
5.3	Application of SDNR numbers with multiplication or division	199
5.4	Parallel addition/subtraction with different representation of addends	200
5.5	SRT division	209

6.	Calculation of special functions	218
6.1	Calculation of logarithms	218
6.2	Calculation of arc tan (y/x)	223
6.3	Calculation of $\sqrt{y/x}$	224
6.4	Inverse functions	226
6.5	CORDIC method for calculating arithmetic functions	228
7.	Time complexity of arithmetic operations	244
7.1	Description of the model	244
7.2	Lower limits of cycle time for arithmetic operations	248
7.3	Upper limits	256
7.4	Calculation of functions Φ_1 and Φ_2 (addition) with binary digit coding of the in/output of the circuit	258

Bibliography 266

Index 273

Symbols 278

1. Number representations

1.1 denary digit codings

Before dealing with digit codings we will define the term "number representation":

Definition 1.1. A *number representation* (or a *notation*) $w^{(n,m)}$ is a mapping which assigns to a sequence $\alpha_{n-1}\ldots\alpha_{-m}$ of digits a value from a range Q (e.g. N, Z, Q).

If $w^{(n,m)}(\alpha_{n-1}\ldots\alpha_{-m}) = q \in Q$, then q is represented by the digit sequence $\alpha_{n-1}\ldots\alpha_{-m}$.

The representation is redundant if there is at least one $q \in Q$ which may be represented in several ways, i.e. if these are sequences

$$\alpha_{n-1}\ldots\alpha_{-m} \neq \beta_{n-1}\ldots\beta_{-m} \quad \text{with:}$$

$$w^{(n,m)}(\alpha_{n-1}\ldots\alpha_{-m}) = w^{(n,m)}(\beta_{n-1}\ldots\beta_{-m}).$$

If there is no ambiguity, either or both of the above w indices are disregarded, i.e. $w = w^{(n)} = w^{(n,m)}$.

Within the scope of this book we will choose as the basic range $Q = \mathbb{Q}$. Real and complex numbers need not be treated separately, since they can be approximated by rational numbers or pairs of rational numbers. We will limit ourselves in the following to digit codings (for more general notations see for example [Ga2]).

Definition 1.2. Let $B_d := \{0, \ldots, d-1\}$. A notation

$$w^{(n,m)} : \{0, d-1\} \overset{\cdot}{\times} \underbrace{B_d \overset{\cdot}{\times} \ldots \overset{\cdot}{\times} B_d}_{(n+m)} \to Q$$

is called *denary digit coding* (with α_{n-1} as sign), if

1. $w(\alpha_{n-1}\ldots\alpha_{-m}) = f(\alpha_{n-1}, \sum_{i=-m}^{n-2} \alpha_i \cdot d^i)$

where $f(\alpha, \gamma_1) \neq f(\alpha, \gamma_2)$ for $\gamma_1 \neq \gamma_2$;

2. $w(\alpha_{n-1}\ldots\alpha_{-m}) \begin{cases} \geq 0 \text{ if } \alpha_{n-1} = 0 \\ \leq 0 \text{ if } \alpha_{n-1} \neq 0 \end{cases}$

For uniqueness of representation of a positive rational number with base d the denary digit codes are not redundant (apart from representation of the zero digit).

The most important characteristics of three denary digit codings are set out below.

Representation by amount and sign (A+S)

$$w_{A+S}(\alpha_{n-1}\ldots\alpha_{-m}) := \begin{cases} \sum_{i=-m}^{n-2} \alpha_i d^i & \text{if } \alpha_{n-1} = 0 \\ -\sum_{i=-m}^{n-2} \alpha_i d^i & \text{if } \alpha_{n-1} = d-1 \end{cases}$$

The zero notation is redundant: $w_{A+S}(d-1\ 0\ldots0) = w_{A+S}(0\ 0\ldots0) = 0$.
The range represented $-d^{n-1}+d^{-m} \leq q \leq d^{n-1}-d^{-m}$ is symmetrical. Coding of $-q$ is obtained from that of $+q$ by changing the sign.

d-complement representation

$$w_d(\alpha_{n-1}\ldots\alpha_{-m}) := \begin{cases} \sum_{i=-m}^{n-2} \alpha_i d^i & \text{when } \alpha_{n-1} = 0 \\ \sum_{i=-m}^{n-2} \alpha_i d^i - d^{n-1} & \text{when } \alpha_{n-1} = d-1 \end{cases}$$

Characteristics:

1. $w_d(\alpha_{n-1}\ldots\alpha_{-m}) < 0$, if $\alpha_{n-1} = d-1$.

The representation is <u>non-redundant</u> since there is only one possible zero code ($w_d(0\ldots0) = 0$).

2. The range represented by $-d^{n-1} \leq q \leq d^{n-1}-d^{-m}$ is asymmetrical.

3. For $\bar{x} := d-1-x$ ($x \in B_d = \{0,1,\ldots,d-1\}$) we have:

$$w_d(\alpha_{n-1}\ldots\alpha_{-m}) + w_d(\bar{\alpha}_{n-1}\ldots\bar{\alpha}_{-m}) = \sum_{i=-m}^{n-2}(d-1)d^i - d^{n-1} = -d^{-m};$$

i.e. $-w_d(\alpha_{n-1}\ldots\alpha_{-m}) = w_d(\bar{\alpha}_{n-1}\ldots\bar{\alpha}_{-m}) + d^{-m}$

$ = w_d(\bar{\alpha}_{n-1}\ldots\bar{\alpha}_{-m}) + w_d(0\ldots0\ 1) = w_d(\bar{\alpha}_{n-1}\ldots\bar{\alpha}_{-m} + 0\ldots0\ 1)$.

The last equation applies only where $(\alpha_{n-1}\ldots\alpha_{-m}) \neq (d-1\ 0\ldots0)$.

If $\alpha_{n-1}\ldots\alpha_{-m}$ is the representation of q $(q \neq -d^{n-1})$, the representation of $-q$ is found by inverting all α_i digits and adding a 1 at the least significant position.

Example: $-w_{10}(9340200) = w_{10}(0659799 + 0000001) = w_{10}(0659800)$.

(d-1)-complement representation

$$w_{d-1}(\alpha_{n-1}\ldots\alpha_{-m}) := \begin{cases} \sum_{i=-m}^{n-2} \alpha_i d^i & \text{when } \alpha_{n-1} = 0 \\ \sum_{i=-m}^{n-2} \alpha_i d^i - (d^{n-1} - d^{-m}) & \text{when } \alpha_{n-1} = d-1 \end{cases}$$

Characteristics:

1. $w_{d-1}(0\ldots0) = w_{d-1}(1\ldots1) = 0$.

2. $w_{d-1}(\bar{\alpha}_{n-1}\ldots\bar{\alpha}_{-m}) + w_{d-1}(\alpha_{n-1}\ldots\alpha_{-m}) = 0$;

i.e. $-w_{d-1}(\alpha_{n-1}\ldots\alpha_{-m}) = w_{d-1}(\bar{\alpha}_{n-1}\ldots\bar{\alpha}_{-m})$.

The representation of $-q$ is obtained by inverting all α_i digits of the $+q$ code.

3. The range represented is symmetrical.

1.2 Insertion of longer notations; overflow problem

The representations treated so far can be inserted because of

$w_A^{(n,m)}(\alpha_{n-1}\ldots\alpha_{-m})$

$= \begin{cases} w_A^{(n+k,m)}(\underbrace{\alpha_{n-1}\ 0\ \ldots\ 0}_{}\ \alpha_{n-2}\ldots\alpha_{-m}) & \text{when } A = \text{'A+S'} \\ w_A^{(n+k,m)}(\underbrace{\alpha_{n-1}\alpha_{n-1}\ldots\alpha_{n-1}}_{k}\alpha_{n-2}\ldots\alpha_{-m}) & \text{when } A = d \text{ or } A = d-1 \end{cases}$

in notations of greater length (hence with larger range or representation).

<u>Definition 1.3.</u> Let $Q_A^{(n,m)} := \{x \in Q \mid w_A^{(n,m)}(\alpha_{n-1}\ldots\alpha_{-m}) = x\}$;
$\min(Q_A^{(n,m)}) := \min\{x \mid x \in Q_A^{(n,m)}\}$; $\max(Q_A^{(n,m)}) := \max\{x \mid x \in Q_A^{(n,m)}\}$.
An arithmetic operation* yields an <u>overflow</u>, when:

$$a*b \notin [\min(Q_A^{(n,m)}), \max(Q_A^{(n,m)})] \quad (a,b \in Q_A^{(n,m)}).$$

If $a*b \notin Q_A^{(n,m)}$, but $a*b \in [\min(Q_A^{(n,m)}), \max(Q_A^{(n,m)})]$, (for example with multiplication or division), a correction can be made by rounding to a $Q_A^{(n,m)}$ number (see for example [Wil]). A clear overflow results in a fault stop or in a faulty interpretation of the result of the operation.

<u>Lemma 1.1.</u> $a,b \in Q_A^{(n,m)} \Rightarrow a \pm b \in Q_A^{(n+k,m)}$ for all $k \geq 1$.

<u>Proof.</u> (1) <u>Amount and sign:</u>

$a,b \in Q_{A+S}^{(n,m)} \Rightarrow 0 \leq |a|, |b| \leq d^{n-1} - d^{-m}$
$\Rightarrow |a \pm b| \leq 2(d^{n-1} - d^{-m}) \leq d^n - d^{-m+1} \Rightarrow a \pm b \in Q_{A+S}^{(n+1,m)}$.

(2) <u>d-complement (or (d-1)-complement):</u>

$$a,b \in Q_A^{(n,m)} \Rightarrow -d^{n-1} \overset{(<)}{\leq} a,b \leq d^{n-1}-d^{-m} \Rightarrow -2d^{n-1} \leq a \pm b \leq 2(d^{n-1}-d^{-m}) \overset{(<)}{}$$

$\Rightarrow -d^n \overset{(<)}{\leq} a \pm b \leq d^n - d^{-m+1} \Rightarrow a \pm b \in Q_A^{(n+1,m)}$.

Lemma 1.1 yields a simple overflow criterion for addition and for subtraction.

<u>Proposition 1.2.</u> a) <u>(overflow recognition in the d- or (d-1)-complement by doubling the sign)</u>

Let $\left.\begin{array}{l} w(\alpha_n \alpha_{n-1}\ldots\alpha_{-m}) \\ w(\beta_n \beta_{n-1}\ldots\beta_{-m}) \end{array}\right\} \in Q_A^{(n,m)}$, i.e. $\alpha_{n-1} = \alpha_n$, $\beta_{n-1} = \beta_n$.

Let $w(\gamma_n \gamma_{n-1}\ldots\gamma_{-m}) = w(\alpha_n\ldots\alpha_{-m}) \pm w(\beta_n\ldots\beta_{-m}) \in Q_A^{(n+1,m)}$.

An overflow (in respect of $Q_A^{(n,m)}$) is recognized by $\gamma_n \neq \gamma_{n-1}$.

(b) *(overflow recognition with amount and sign)*

Let $\left. \begin{array}{l} w(\alpha_n \ 0 \ \alpha_{n-2}\ldots\alpha_{-m}) \\ w(\beta_n \ 0 \ \beta_{n-2}\ldots\beta_{-m}) \end{array} \right\} \in Q_{A+S}^{(n,m)}$

$w(\gamma_n\gamma_{n-1}\ldots\ldots\gamma_{-m}) = w(\alpha_n\ldots\alpha_{-m}) \pm w(\beta_n\ldots\beta_{-m}) \in Q_{A+S}^{(n+1,m)}$.

An overflow (in respect of $Q_{A+S}^{(n,m)}$) is recognized by $\gamma_{n-1} \neq 0$.

1.3 Arithmetic for denary digit systems

1.3.1 Formal sum of two denary numbers

Definition 1.4. Let $\alpha = \alpha_{n-1}\ldots\alpha_{-m}$, $\beta = \beta_{n-1}\ldots\beta_{-m}$ be two denary digit numbers ($\alpha_i, \beta_i \in B_d$, $\alpha_{n-1}, \beta_{n-1} \in \{0, d-1\}$).

$\gamma := \gamma_n\gamma_{n-1}\ldots\gamma_{-m}$ means *formal sum* of α and β ($\gamma = \alpha+\beta$),

where $\sum_{i=-m}^{n-1} \alpha_i d^i + \sum_{i=-m}^{n-1} \beta_i d^i = \sum_{i=-m}^{n} \gamma_i d^i$ ($\gamma_i \in B_d$).

The formal sum of two denary numbers is uniquely defined.

$\gamma_n \in \{0,1\}$ means *total carry*.

1.3.2 d-complement addition

Proposition 1.3. Let $\alpha = \alpha_{n-1}\ldots\alpha_{-m}$, $\beta = \beta_{n-1}\ldots\beta_{-m}$,
$\gamma = \gamma_n\gamma_{n-1}\ldots\gamma_{-m} = \alpha+\beta$.

If $w_d(\alpha) + w_d(\beta) \in Q_d^{(n,m)}$ (i.e., if there is no overflow situation), then we have:

$w_d(\gamma_{n-1}\ldots\gamma_{-m}) = w_d(\alpha) + w_d(\beta)$.

Proof. We will insert α and β in longer representations:

$\alpha' := \alpha_{n+k-1}\ldots\alpha_n\alpha_{n-1}\ldots\alpha_{-m}$
$\beta' := \beta_{n+k+1}\ldots\beta_n\beta_{n-1}\ldots\beta_{-m}$ $\left[k \geq 0, \begin{array}{l} \alpha_{n+k-1} = \ldots = \alpha_{n-1} \\ \beta_{n+k-1} = \ldots = \beta_{n-1} \end{array} \right]$

Since $w_d^{(n+k)}(\alpha') = w_d^{(n)}(\alpha)$ then:

$w_d^{(n+k)}(\alpha') + w_d^{(n+k)}(\beta') = w_d^{(n)}(\alpha) + w_d^{(n)}(\beta) \in Q_d^{(n,m)}$.

For the formal sum $\gamma' = \gamma_{n+k}\gamma_{n+k-1}\cdots\gamma_n\gamma_{n-1}\cdots\gamma_{-m} = \alpha' + \beta'$

hence $\gamma_{n-1} = \gamma_n = \gamma_{n+1} = \cdots = \gamma_{n+k-1}$,

i.e. $w_d^{(n+k)}(\gamma_{n+k-1}\cdots\gamma_{-m}) = w_d^{(n)}(\gamma_{n-1}\cdots\gamma_{-m})$.

Furthermore:

$w_d^{(n+k)}(\alpha_{n+k-1}\cdots\alpha_{-m}) + w_d^{(n+k)}(\beta_{n+k-1}\cdots\beta_{-m})$

$\overset{(*)}{=} [w_d^{(n+k+1)}(0\ \alpha_{n+k-1}\cdots\alpha_{-m}) + w_d^{(n+k+1)}(0\ \beta_{n+k-1}\cdots\beta_{-m})] \bmod d^{n+k}$

$= \sum_{i=-m}^{n+k} \gamma_i d^i \bmod d^{n+k} = \sum_{i=-m}^{n+k-1} \gamma_i d^i \bmod d^{n+k}$

$= w_d^{(n+k+1)}(0\ \gamma_{n+k-1}\cdots\gamma_{-m}) \bmod d^{n+k}$

$\overset{(*)}{=} w_d^{(n+k)}(\gamma_{n+k-1}\cdots\gamma_{-m}) \bmod d^{n+k}$

$= w_d^{(n)}(\gamma_{n-1}\cdots\gamma_{-m}) \bmod d^{n+k}$

$= w_d^{(n)}(\gamma_{n-1}\cdots\gamma_{-m})$ (since $k \geq 0$ arbitrary)

The equations (*) apply since:

$w_d^{(n+1)}(0\ \gamma_{n-1}\cdots\gamma_{-m}) = w_d^{(n)}(\gamma_{n-1}\cdots\gamma_{-m}) \bmod d^n$.

The proposition is thus proved.

d-complement addition with overflow recognition:	
$\alpha = \alpha_{n-1}\alpha_{n-1}\alpha_{n-2}\cdots\alpha_{-m}$	$\gamma_n \neq \gamma_{n-1} \Leftrightarrow$ overflow in respect of $Q_d^{(n,m)}$;
$\beta = \beta_{n-1}\beta_{n-1}\beta_{n-2}\cdots\beta_{-m}$	$\gamma_n = \gamma_{n-1} \Leftrightarrow w_d(\alpha) + w_d(\beta)$
$\gamma = \gamma_n\ \gamma_{n-1}\gamma_{n-2}\cdots\gamma_{-m}$	$= w_d(\gamma_{n-1}\cdots\gamma_{-m})$.

Examples: $d = 10$.

1) $\alpha =\ \ 993278\ \ \triangleq -\ 6722$ 2) $\alpha =\ \ 995213\ \ \triangleq -\ 4787$

      $\beta =\ \ 007945\ \ \triangleq +\ 7945$ $\beta =\ \ 993174\ \ \triangleq -\ 6826$

      $\gamma = \mathit{1}001223\ \ \triangleq +\ 1223$ $\gamma = \mathit{1}988387\ \ \triangleq -11613$

        no overflow   overflow in respect of $Q^{(5,0)}$
                                               no overflow in respect of $Q^{(6,0)}$.

1.3.3 (d-1)-complement addition

As in the d-complement, the (d-1)-complement addition is effected by formal summation of the two addends.

The total carry has to be added to the least significant position (end-around-carry, calculation of a further formal sum). No new total carry results from this; the proof of this method is given in the following:

Proposition 1.4. Let $\alpha = \alpha_{n-1}..\alpha_{-m}$, $\beta = \beta_{n-1}..\beta_{-m}$, $\gamma = \gamma_n \gamma_{n-1}..\gamma_{-m} = \alpha + \beta$.
If $w_{d-1}(\alpha) + w_{d-1}(\beta) \in Q_{d-1}^{(n,m)}$, we have

a) $\gamma_{n-1}...\gamma_{-m} + 0\,0...0\,\gamma_n = \gamma_n^* \gamma_{n-1}^* ...\gamma_{-m}^*$ with $\gamma_n^* = 0$.

b) $w_{d-1}(\gamma_{n-1}^*...\gamma_{-m}^*) = w_{d-1}(\alpha) + w_{d-1}(\beta)$.

Proof. a) hypothesis: $\gamma_n^* \neq 0 \Rightarrow \gamma_n = 1$; $\gamma_{n-1} = ... = \gamma_{-m} = d-1$

$$\Rightarrow \sum_{i=-m}^{n-1} \alpha_i d^i + \sum_{i=-m}^{n-1} \beta_i d^i \leq 2(d^n - d^{-m}) \leq 2 \times d^n - d^{-m} = \sum_{i=-m}^{n} \gamma_i d^i$$

$\Rightarrow \alpha + \beta \neq \gamma$ (contradiction).

b) Using the signs of proposition 1.3, we have:

$$w_{d-1}^{(n)}(\alpha) + w_{d-1}^{(n)}(\beta) = w_{d-1}^{(n+k)}(\alpha') + w_{d-1}^{(n+k)}(\beta')$$

$$= [w_{d-1}^{(n+k+1)}(0\alpha_{n+k-1}\,\alpha_{-m}) + w_{d-1}^{(n+k+1)}(0\beta_{n+k-1}..\beta_{-m})] \bmod (d^{n+k} - d^{-m})$$

$$= \sum_{i=-m}^{n+k} \gamma_i d^i \bmod (d^{n+k} - d^{-m})$$

$$= [\sum_{i=-m}^{n+k-1} \gamma_i d^i + \gamma_{n+k} d^{-m}] \bmod (d^{n+k} - d^{-m})$$

$$= w_{d-1}^{(n+k)} \begin{bmatrix} \gamma_{n+k-1}...\gamma_{-m} \\ +0\quad\quad ...\gamma_{n+k} \end{bmatrix} \bmod (d^{n+k} - d^{-m})$$

$$= w_{d-1}^{(n)} \begin{bmatrix} \gamma_{n-1}...\gamma_{-m} \\ +\,0\,\,...\gamma_n \end{bmatrix} \bmod (d^{n+k} - d^{-m}) = w_{d-1}^{(n)} \begin{bmatrix} \gamma_{n-1}\,...\gamma_{-m} \\ +\,0\,\,\,...\gamma_n \end{bmatrix}$$

$$= w_{d-1}^{(n)}(\gamma_{n-1}^*...\gamma_{-m}^*) \;.$$

(d-1)-complement addition with overflow recognition:

$\alpha =$	$\alpha_{n-1}\alpha_{n-1}\alpha_{n-2}\ldots\alpha_{-m}$	$\gamma_n^* \neq \gamma_{n-1}^* \Leftrightarrow$ overflow in respect of $Q_{d-1}^{(n,m)}$;
$\beta =$	$\beta_{n-1}\beta_{n-1}\beta_{n-2}\ldots\beta_{-m}$	$\gamma_n^* = \gamma_{n-1}^* \Leftrightarrow w_{d-1}(\alpha) + w_{d-1}(\beta)$
$\gamma =$	$\gamma_{n+1}\ \gamma_n\ \gamma_{n-1}\gamma_{n-2}\ldots\gamma_{-m}$	$= w_{d-1}(\gamma_{n-1}^* \ldots \gamma_{-m}^*)$.
	(+; end around carry)	
$\gamma^* =$	$\gamma_n^*\ \gamma_{n-1}^*\gamma_{n-2}^*\ldots\gamma_{-m}^*$	

Examples: $d = 10$.

1) $\alpha = 993278 \triangleq -6721$
   $\beta = 007951 \triangleq +7951$
   $\gamma = 1001229$ ↵
   $\gamma^* = 001230 \triangleq +1230$
   no overflow

2) $\alpha = 005413$
   $\beta = 008179$
   $\gamma = 013592$
   overflow in respect of $Q_{d-1}^{(5,0)}$
   no overflow in respect of $Q_{d-1}^{(6,0)}$.

1.3.4 Addition with notation by amount and sign

Let $\alpha = \alpha_{n-1}\ 0\ \alpha_{n-2}\ldots\alpha_{-m}$ and $\beta = \beta_{n-1}\ 0\ \beta_{n-2}\ldots\beta_{-m}$ be two addends represented by A+S. We are seeking a method for the calculation of $\gamma = \gamma_n\gamma_{n-1}\gamma_{n-2}\ldots\gamma_{-m}$ with $w_{A+S}(\gamma) = w_{A+S}(\alpha) + w_{A+S}(\beta)$. An obvious method (see, for example, [Hol]) is to add the "amounts" $\alpha_{n-2}\ldots\alpha_{-m}$ and $\beta_{n-2}\ldots\beta_{-m}$, if $\alpha_{n-1} = \beta_{n-1}$, and subtract the smaller from the larger amount if $\alpha_{n-1} \neq \beta_{n-1}$. Overflow can occur only in the first case; this is indicated by $\gamma_{n-1} \neq 0$. Apart from some exceptions, this method requires calculation of the maximum amount (with addends of different signs).

A more suitable method (see 1.3.5) is first to transform both addends to the (d-1)-complement, carry out addition there and convert the result again into amount and sign.

1.3.5 Transition to another notation

The denary digit codings we have dealt with differ only in representation of negative numbers. We have:

1. $w_{A+S}\,[d-1\ \alpha_{n-2}..\alpha_{-m}] = w_{d-1}\,[d-1\ \bar{\alpha}_{n-2}..\bar{\alpha}_{-m}] = w_d\,\left[d-1\ \bar{\alpha}_{n-2}..\bar{\alpha}_{-m\atop +1}\right]$

2. $w_{d-1}\,[d-1\ \alpha_{n-2}..\alpha_{-m}] = w_{A+S}\,[d-1\ \bar{\alpha}_{n-2}..\bar{\alpha}_{-m}] = w_d\,\left[d-1\ \alpha_{n-2}..\alpha_{-m\atop +1}\right]$

3. $w_d\,[d-1\ \alpha_{n-2}..\alpha_{-m}] = w_{d-1}\,\left[d-1\ \alpha_{n-2}..\alpha_{-m\atop -1}\right] = w_{A+S}\,\left[d-1\ \bar{\alpha}_{n-2}..\bar{\alpha}_{-m\atop +1}\right]$

(when $d-1\ \alpha_{n-2}...\alpha_{-m} \neq d-1\ 0...0$) .

The same range of representation for the (d-1)-complement or for A+S and simple transformation method between these two codings is provided by the following A+S -addition method:

```
Addition of two A+S-addends:
```

$$\left.\begin{array}{l}\alpha = \alpha_{n-1}.......\alpha_{-m} \\ \beta = \beta_{n-1}.......\beta_{-m}\end{array}\right] \to \begin{array}{l}\alpha^* = \alpha_{n-1}\alpha^*_{n-2}.....\alpha^*_{-m} \\ \beta^* = \beta_{n-1}\beta^*_{n-2}.....\beta^*_{-m}\end{array}$$

$(d-1)$-complement addition

$$\varepsilon = \varepsilon_n \varepsilon_{n-1}\varepsilon_{n-2}.....\varepsilon_{-m}$$
(end around carry)

$$\delta_{n-1}\delta^*_{n-2}...\delta^*_{-m} \leftarrow \delta_{n-1}\delta_{n-2}.....\delta_{-m}$$

where $\alpha^*_i := \begin{cases}\alpha_i; & \text{where } \alpha_{n-1} = 0 \\ d-1-\alpha_i; & \text{where } \alpha_{n-1} = d-1\end{cases}$ (β^*_i, δ^*_i analog) .

If no overflow has occurred (recognized by extension of the coding by one additional place), we have:

$$w_{A+S}(\alpha) + w_{A+S}(\beta) = w_{A+S}(\delta_{n-1}\delta^*_{n-2}.......\delta^*_{-m}) .$$

Examples: $d = 10$.

```
1) -7681 | α= 907681⎤  992318    2) -1248 | α= 901248⎤  998751
   +2435 | β= 002435⎦→ 002435       +4358 | β= 004358⎦→ 004358
                      ────────                         ────────
                       0994753                          1003109
                      └──────↑                         └──────↑

   ─────   ───────────────────       ─────   ───────────────────
   -5246    905246 ← 994753           +3110    003110 ← 003110
```

1.4 Other representations

Apart from the denary digit codings treated so far, other methods of representation have been examined which might be of advantage for special applications. These less usual representations mainly have the great disadvantage that their application to operations for which they are not adapted is extraordinarily costly and time consuming.

1.4.1 Redundant notations (see 5.1-5.3)

We are here concerned with redundant digit coding for which negative α_i numbers are also possible. A carry for addition (subtraction) is picked up by the redundants at the next position, and cannot therefore continue to all positions, as for the denary digit codings dealt with at 1.1. The range of addition (and therefore the range of operations attributable to addition, for example multiplication) is thus reduced considerably on average. Greater effort is necessary for overflow recognition and for deconversion (changing from redundant to non-redundant representation).

There are algorithms, for example the SRT-division (see 5.5), which combine the advantage of denary digit coding with the advantage of redundant notation.

1.4.2 Residue arithmetic ([Ga1], [Ba1], [Ba2], [Sa3])

In this type of representation, numbers are coded by their residue classes in accordance with various prime number modules. Hence the addition (subtraction) or multiplication time is shorter as compared with digit systems dealt with so far, since only the (considerably smaller) residue classes are to be added or multiplied. This can be performed parallel for all residue classes. This time saving, however, has to be paid for at high price, since determination of sign, division, detection of overflow situations, scaling and conversion or deconversion are very costly. It is doubtful whether the known algorithms for solution of these problems can be essentially improved (i.e. speeded up and/or cheapened). If this is impossible, residue arithmetic is merited only for a very limited class of tasks (such as problems for which the sign of all operands and results is known, which furthermore are free of division and for which it is known at the start that overflow situations cannot intervene).

1.5 Basic choice, register, connections, microprograms

1.5.1 Optimum basic choice

The operands of arithmetic operations, their results and intermediate results in the case of complicated operations are stored in registers of a fixed word length. For a denary register R of length n, the following representation is employed:

$$R = [R_{n-1}, \ldots, R_0] \triangleq \boxed{R_{n-1} \mid R_{n-2} \mid \ldots \mid R_0} \qquad (R_i \in B_d) .$$

R can have d^n different states, i.e. it can be used to present up to a maximum of d^n different numbers. The maximum state number is reduced to $2d^{n-1}$, where R_{n-1} is interpreted as sign.

Lemma 1.5. *If the costs C of a register are proportional to the length n and to the basis d, i.e. C = const · d · n, then the costs will be least for representation of M numbers (M → ∞) for the base d = e = 2.71...*

Of the integer base, the optimal is d = 3. The costs for d = 2 and d = 4 are the same and only slightly higher than for d = 3.

Proof. For large M we have approximately: $d^n = M$, i.e. $n = \frac{\log M}{\log d}$.
The total costs $C = C(d) = \text{const} \cdot d \cdot \frac{\log M}{\log d}$ become minimal for $d = e$.
In practice, binary registers are used exclusively at present. Apart from technical limitations (other registers not available are far too costly), the reliability aspect is important when choosing. In general, denary connections are more susceptible to trouble than binary. Hence, denary digits must be coded by a series of binary digits; this is possible without loss only for $d = 2^k$ ($k \in \mathbb{N}$). This is why nearly all algorithms for arithmetic operations are based on a representation of operands and results on the basis $d = 2^k$.

1.5.2 Realisation of arithmetic algorithms, microprograms

Algorithms for arithmetic operations are obtained by circuits or connections. The main purpose of this book is to describe efficient algorithms. We cannot go into details with regard to the control of circuits or connections (see for example [Hu1], [Le3], [K11]).

Connections can be described very clearly by microprograms. Each line of a microprogram corresponds to one function of the connection. In the same way as an ALGOL-type notation, it contains all operations performed by the connection in this function (conditional register transfers; changing of counter positions; allocation of constants to registers etc.). All instructions of a microprogram line are performed in parallel. This means that the line instructions must be unambiguous, for example multiple assignments to a register position are prohibited if they are not mutually exclusive. Furthermore, a microprogram line contains a reference to the line which includes switching operations to be carried out in the next function. Most microprograms contain loops (multiple cycling of various functions). Special register contents and counter positions can be taken as criterion for leaving a loop or for finishing a microprogram.

1.6 Representation of fixed and floating points

1.6.1 Fixed point

We designate a denary digit coding $w^{(n,m)}$ (see definition 1.2) also as a fixed-point representation (point after the n-th digit from the left). The codings $w^{(n,0)}$ (point after the last digit, representation of <u>integral</u> numbers) and $w^{(1,n-1)}$ (representation of rational numbers from the interval [-1:1]) are often used in a register of length n. Changing the position of a point is designated as <u>scaling</u>; it corresponds to multiplication by a power base d and can be performed by shifting the register to the right or to the left. Numbers which cannot be registered in a fixed-point register must be scaled first; this is performed automatically with the floating-point representation.

1.6.1.1 <u>Overflow and scaling problems with fixed-point representation</u>

1. Addition or subtraction

Overflow situations can be recognized by an additional register position and can be removed by scaling. Scaling performed on operands influences the sequence and result of arithmetic operations. In fig. 1.1, the addend c must be scaled because it is added to a scaled addend (a+b). Furthermore, scaling gives rise to errors of rounding if the digits dropped from a register are not stored and included in the arithmetic. These problems gave rise to the introduction of a floating-point arithmetic.

2. Multiplication (fig. 1.2)

The product of two n-digit numbers has a length of 2n. The register $MP^{(1)}$ contains the most significant digits of the product.

<u>Integral</u> multiplication yields an overflow if the product cannot be located in the register $MP^{(2)}$. Assuming that the factors are distributed uniformly over the representation range, on average every second multiplication ends with an overflow.

Fig. 1.1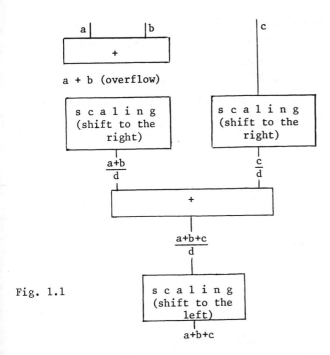

Fig. 1.2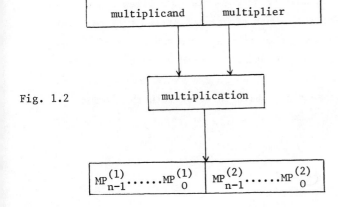

Overflow cannot occur when applying the notation $w^{(1,n-1)}$. In order to achieve greater accuracy, it may be useful to continue to calculate with the product twice the length; for example, if a product occurs as a dividend in a division. The product is reduced to a single length by omitting $MP^{(2)}$ and rounding $MP^{(1)}$ (correction of the least significant digit of $MP^{(1)}$ as a function of $MP^{(2)}_{n-1}$; see also 3.1).

3. Division

Many differentiations and scalings of the dividend and/or the divisor are necessary in order to calculate the quotient with sufficient accuracy. The work involved in scaling depends on the notation chosen.

Most algorithms (see chapter 4) proceed from the notation $w^{(1,n-1)}$. If the dividend DD and the divisor DR are scaled in such a manner, that $0 \leq |w(DD)| < |w(DR)| \leq 1$, we have $0 \leq \left|\frac{w(DD)}{w(DR)}\right| < 1$; hence the quotient lies within the same area as the dividend and the divisor.

1.6.1.2 Distribution of the represented numbers

The difference between two adjacent represented fixed-point numbers is constant (equal distribution over the range of representation). From experience, numbers of smaller amounts are more often employed than numbers of larger amounts. Most programmers attempt to restrict the size of the numbers occurring or to use a greater range of representation (double precision) in order to avoid an overflow situation. Hence, a representation is of advantage where the separations between the adjacent represented numbers increase with the amount of numbers (floating-point arithmetic).

1.6.2 Floating-point arithmetic representation

1.6.2.1 Definition and rules of calculation

Floating-point calculation was introduced around 1940 for automatic treatment of the scaling problems occurring in connection with fixed-point operands or results.

Definition 1.5. *1. (a,e,d) signifies (non-normalized) floating-point representation* of $z \in R$ *with base d, if* $z = a \cdot d^e$ *($e \in Z$). a is designated mantissa and e the exponent.*

2. (a,e,d) signifies normalized floating-point representation of $z \in R$.

$$\Leftrightarrow z = \begin{cases} a \cdot d^e & e \in Z, \frac{1}{d} \leq |a| < 1 \quad \text{where } z \neq 0 \\ 0 \cdot d^e & \text{where } z = 0. \end{cases}$$

Rules of calculation for denary floating-point numbers:

1. $(a_1, e_1, d) \cdot (a_2, e_2, d) = (a_1 \cdot a_2, e_1 + e_2, d)$;

2. $(a_1, e_1, d) / (a_2, e_2, d) = (a_1 / a_2, e_1 - e_2, d)$;

3. $(a_1, e_1, d) \pm (a_2, e_2, d) = \begin{cases} (a_1 \pm a_2 \cdot d^{-(e_1 - e_2)}, e_1, d) & \text{where } e_1 \geq e_2 \\ (a_1 \cdot d^{-(e_2 - e_1)} \pm a_2, e_2, d) & \text{where } e_1 < e_2. \end{cases}$

When adding or subtracting, the exponents must first be matched by shifting the mantissa of the addend with the smaller exponent $|e_1 - e_2|$ to the right (preshifts). Scaling expenditure for this operation and rounding faults are higher than for multiplication or division (see [Wil]).

1.6.2.2 Normalisation

Automatic scaling with simultaneous rounding faults control is possible only in normalized representation. While the number of forms of non-normalized floating-points for $z \in R$ are infinite, the normalized representation for $z \neq 0$ is uniquely determined.

Mantissa and exponent of a normalized floating-point number are usually stored in their own n or h length registers. Generally speaking, h (exponent length) is much smaller than n; if n length registers only are available, the free positions of the exponent register can be used for increasing the word length (i.e. precision of the mantissa), (see 1.6.3.3).

The codings $w^{(1,n-1)}$ or $w^{(h,0)}$ are generally used for mantissa or exponent notations; the choice of (d-1)-complement, d-complement or representation by amount and sign is dependent on the configuration of the machine and the arithmetic operation to be performed.

According to these assumptions we have:

1. $\frac{1}{d} \le |a| \le 1 - \frac{1}{d^{n-1}}$ for all digit codings treated by us.

Remark: $a=-1$ is possible in the d-complement; on account of $|a| \stackrel{!}{<} 1$, this value is normalized to $a = \frac{1}{d}$.

2. $e \in E := [e_{min} : e_{max}]$ with $e_{min}, e_{max} \in Z$.

It is desired that each integral number of the interval E can be represented in a register of length h,

i.e. $e_{max} - e_{min} \le 2 \cdot d^{h-1} - 1$

(the first position of the register can have only two different values).

1.6.2.2.1 Prenormalisation; exponent overflow and underflow

On account of $(a,e,d) = (a \cdot d, e - 1, d) = (a/d, e+1, d)$, each floating-point number $z \triangleq (a,e,d)$ with $z \neq 0$ can be normalized by a sequence of denary shifts of the mantissa to the left (right) and simultaneous lowering (raising) of the exponent, if the exponent range is unrestricted. Restriction of the exponent to the interval $[e_{min} : e_{max}]$ can have two effects:

1. **Exponent overflow**

There is an exponent overflow, if:

$$z \triangleq (a,e,d) \text{ with } |a| \geq 1, e = e_{max}.$$

Within the limits of the range of notations, the floating-point number can no longer be normalized: fault stop $z := \infty$.

2. **Exponent underflow**

$$z \triangleq (a,e,d) \text{ with } |a| < \frac{1}{d}, e = e_{min}, \text{ i.e. } |z| < d^{e_{min}-1}.$$

The z quantity is smaller than that of any other representable number (except 0). Hence, if $z \neq 0$, z can be regarded as an infinitely small quantity ε with $\text{sign}(\varepsilon) = \text{sign}(z)$. For ε and the quantity ∞ which occurs in an overflow, inherent rules for calculation are appropriate (see, for example [Bu1]). Another simple solution to the overflow problem is rounding to the smallest adjacent number $\neq 0$, i.e. to $z_{min} := (\text{sign}(z) \times \frac{1}{d}, e_{min}, d)$. With each automatic correction of an overflow situation, a message should be passed to the user.

1.6.2.2.2 Post-normalisation, "Dirty zero"

Applying the normalized representation only, the result of an arithmetic operation a*b must be normalized by a series of (post) shifts:

1. **Multiplication and division**

a. $(a_1, e_1, d_1) \cdot (a_2, e_2, d)$

$$= \begin{cases} (a_1 \cdot a_2, e_1 + e_2, d) & \text{where } \frac{1}{d} \leq |a_1 \cdot a_2| < 1 \text{ and } e_1 + e_2 \in E \\ (a_1 \cdot a_2 \cdot d, e_1 + e_2 - 1, d) & \text{where } \frac{1}{d^2} \leq |a_1 \cdot a_2| < \frac{1}{d} \text{ and } e_1 + e_2 - 1 \in E \\ z_{min} & \text{where } e_1 + e_2 < e_{min} \text{ (underflow)} \\ \infty & \text{otherwise (overflow)}. \end{cases}$$

b. $(a_1,e_1,d)/(a_2,e_2,d)$

$$= \begin{cases} (a_1/a_2, e_1-e_2, d) & \text{where } \frac{1}{d} < |a_1/a_2| < 1 \text{ and } e_1-e_2 \in E \\ (a_1/a_2 \cdot \frac{1}{d}, e_1-e_2+1, d) & \text{where } 1 \le |a_1/a_2| < d \text{ and } e_1-e_2+1 \in E \\ \infty & \text{where } e_1-e_2 > e_{max} \quad \text{(overflow)} \\ z_{min} & \text{otherwise (underflow)} . \end{cases}$$

For normalisation (if at all possible), one single post shift is adequate in these two operations.

2. Addition and subtraction

Here, the addend may be mutually cancelling partially or entirely. The number of normalisation shifts depends on the difference in exponents and on the sign of the addends.

Example: $a = (0.99999994, 17, 10) = (0.9999999, 18, 10)$
          $b = (0.10000002, 18, 10)$
          ──────────────────────────
          $a - b = (-0.00000003, 18, 10) = (-0.3, 11, 10)$.

Rounding errors are particularly high with partial cancellation (see [Wil]). Further calculations with such results frequently yield very unprecise and dubious results.

Partial cancellation is particularly effective if the exponents of the two addends are the same and if the addends have opposing (addition) or equal signs (subtraction). In the extreme case, total cancellation occurs ("dirty zero"):

$$(a,e,d) - (a,e,d) = (0,e,d) .$$

This number must not be interpreted simply as zero; all we know concerning this, is the relationship

$$- d^{e-(n-1)} < (0,e,d) < + d^{e-(n-1)} .$$

If $e_{max} \gg n$ and $e \approx e_{max}$, $(0,e,d)$ can be situated within an enormously large range (d=2, n=60, e=128 $\Rightarrow$ $-2^{69} < (0,e,d) < 2^{69}$).

There are many more "dirty zero"-problems, which we cannot deal with (see [Bu1]).

1.6.3 Basic selection in floating-point representations.

Apart from basis d of a floating-point number $z = (a,e,d)$, the bases d_a and d_e of the internal mantissa or exponent representation are to be established.

If mantissa operations (multiplication with d or division by d) can be carried out by shifting the mantissa, we have $d = d_a^r$ ($r \in \mathbb{N}$), hence a denary shift is replaced by r shifts to the base d_a. Since bases of the form $d = 2^k$ ($k \in \mathbb{N}$) are used almost exclusively, this means:

$$d_a = 2^u, \quad d = 2^k = d_a^r \quad (k = u \cdot r,\ k,u,r \in \mathbb{N}).$$

1.6.3.1 Base d_e of the exponent

Changes in exponents (addition or subtraction of smaller integral numbers) are associated with mantissa operations. Usually therefore:

$$d_e = 2 \quad \text{or:} \quad d_e = d_a = 2^u \quad (u \in \mathbb{N}).$$

The notation of the exponent is selected so that nearly as many (and as large) positive as negative exponents can be represented. With a binary register of the length h and with a 2-complement notation, we obtain for E:

$$E = [e_{min} : e_{max}] = [-2^{h-1} : 2^{h-1}-1] \quad \text{(range of exponents)}.$$

1.6.3.2 Base of the mantissa and the floating-point number

The easiest solution is $d = d_a = 2$; in this case, all three bases are the same. However, application of a base $\dot{d} = 2^k$ ($k > 1$) is advantageous for different reasons:

1. Extension of the range of representation:

$$z = a \cdot 2^e \xleftarrow{d = 2} (a,e,d) \xrightarrow{d = 2^k} z' = a \cdot (2^k)^e$$

2. Reduction of the number of shifts through larger shifts:

Only binary shifts are necessary over a fixed number k of bits. The differences in exponents are smaller than for base d = 2 by a factor k.

3. Shorter exponent:

Fewer exponent bits are necessary for covering an area roughly as large as with base d = 2. The free bits can be used either for shortening the register (reduced costs) or for extending the mantissa (increased precision).

A disadvantage of using a base $d = 2^k$ (k > 1) is loss of precision. The gaps between representable numbers and accordingly the rounding errors increase with extension of the representation range. Furthermore, checking of a mantissa on normalisation can be rendered more difficult, since a normalized number can start with as many as k zeros or unities (including the sign).

1.6.3.3 Examples

1. $d = d_a = 2^4$; exponent length h = 7, mantissa length n = 25, $d_e = 2$ (IBM 360)

The mantissa can take on 6 digits on the base d = 16; the first mantissa bit contains the sign.

Negative numbers are represented by amount and sign. This provides a particularly simple check on normalisation. The exponent is on the left of the mantissa; this is preferable for high speed calculation (extension of mantissa):

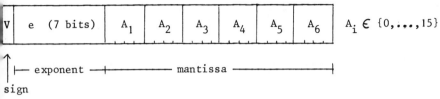

├─ exponent ─┼─────── mantissa ───────┤

↑
│
sign

Range of numbers: $E = [-64:+63]$; $\frac{1}{16} \cdot 16^{-64} \leq |z| \leq (1 - \frac{1}{16^6}) \cdot 16^{+63}$.

2. $d = d_a = d_e = 2$, word length $h+n = 32$

$h = 9$ is necessary to cover roughly the same range of numbers
$2^{-260} \leq |z| \leq (1 - \frac{1}{2^{24}}) \cdot 2^{252}$ as in example 1 .

Hence, the mantissa is shortened by 2 bits which partially compensates the loss of precision through using the base 16 .

| V | e (9 bits) | a_1 ... a_{22} | $a_i \in \{0;1\}$ |

├─ exponent ─┼─────── mantissa ───────┤

Range of numbers: $E = [-256:+255]$; $2^{-257} \leq |z| \leq (1 - \frac{1}{2^{22}}) \cdot 2^{+255}$.

Both configurations have approximately the same range of representation and precision characteristics. Advantages of base $d = 16$ (example 1) over $d = 2$ are fewer preshifts, postshifts and exponent operations.

3. Word structure of the TR 440 computer

This machine operates with a word length of 52 bits; 48 bits accomodate the actual information (mantissa or exponent with floating-point numbers, operation and address sections with instructions, coded text symbols when representing alphanumeric symbols), 2 bits are used for type recognition (differentiation between fixed and floating-point numbers, instructions and alphatext), 2 are test bits (making up the total binary digits of the word to the same residual class modulo 3, i.e. triple checking).

a. Structure of a simple long floating-point number (base $d = 2^4$)

| Pr | 00 | A_1 vv | A_2 | A_3 | A_4 | A_5 | A_6 | A_7 | A_8 | A_9 | A_{10} | v | exponent |

|←——————————— mantissa ———————————→|

↑—— sign

b. Structure of a simple long fixed-point number

| Pr | 01 | vv | a_3 | binary number; 48 bits with sign | a_{48} |

└── type recognition

c. Instruction word (each word contains two instructions)

| Pr | 10 | operation section | address section | operation section | address section |

└── parity bits

d. Alphanumeric symbols

| Pr | 11 | alphatext coding |

2. Adders

2.1 (m,k)-counter, halfadder, fulladder

Definition 2.1. *A denary (m,k)-counter is a circuit with m inputs and k outputs, defined by:*

$$f_{(m,k)} : B_d^m \supset D \to B_d^k$$

$$(a_1,\ldots,a_m) \to (z_{k-1},\ldots,z_0) \text{ with } \sum_{j=1}^{m} a_j = \sum_{i=0}^{k-1} z_i \cdot d^i .$$

$f_{(3,2)}$ *is called* **fulladder**, $f_{(2,2)}$ *is called* **halfadder**.

Conclusion. *1. Let* $u_D := \max\{ \sum_{i=1}^{m} a_i \mid (a_1,\ldots,a_m) \in D\}$. *Hence we have:*

$u_D \le d^k-1$, *i.e.* $k \ge \lceil \log_d(u_D+1) \rceil$.

2. For $D = B_d^m$ *we have:* $u_D = m \cdot (d-1)$, *i.e.* $k \ge \lceil \log_d[m \cdot (d-1)+1] \rceil$.

Example: Let $d = 2$ and $D := B^4 \setminus (1,1,1,1)$. Hence we have $u_D = 3$, i.e. $k \ge 2$. Since the input combination $(1,1,1,1)$ is excluded, a $(4,2)$-counter can be used; this counter is introduced in the form of a special multiplier ([Fe3], see also 3.4.8.2) .

Construction of binary (m,2)-counters (m $\in$ {2,3,4}) over {v,·,⁻} in modular design. Cycle time and cost considerations

$z_0 = s = a_1 \oplus a_2 \oplus \ldots \oplus a_m$ (sum)

$z_1 = c = \lfloor \sum_{i=1}^{m} a_i/2 \rfloor$ (carry)

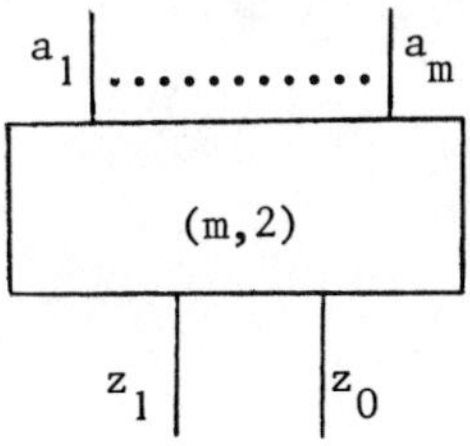

Fig. 2.1

We measure the _expenditure_ κ_{SK} necessary for an SK circuit with the Hotz cost function [Hol], which regards negation as cost free (double rail logic) and which takes the sum of inputs in the "AND" and "OR" type modular units as a measure of the costs. We define _cycle time_ (number of steps) τ_{SK} of a circuit as the maximum number of series AND and OR modular units.

a) (2,2)-counter (halfadder); HA

$$c = a_1 \cdot a_2 ;$$
$$s = a_1 \oplus a_2 = a_1 \bar{a}_2 \vee \bar{a}_1 a_2 \qquad (*)$$
$$= (a_1 \vee a_2) \cdot \overline{a_1 \cdot a_2} = (a_1 \vee a_2) \cdot \bar{c} \qquad (**)$$

Remark: Throughout this manuscript, ab is considered to be equivalent to the notation a·b.

$$\kappa_{HA} = \begin{cases} 8 \text{ calculation of s according to } (*) \\ 6 \text{ calculation of s according to } (**) ; \end{cases}$$

$$\tau_{HA} = 2 .$$

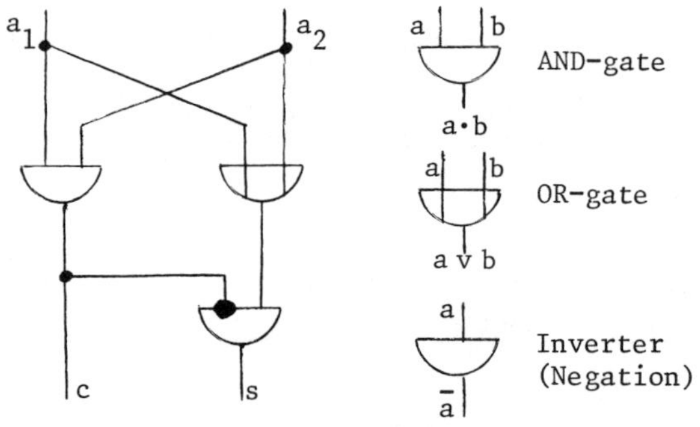

Fig. 2.2
(halfadder)

AND-gate

OR-gate

Inverter
(Negation)

When calculating s according to formula (**) (see fig. 2.2), the carry c logic is applied; the result of this is a reduction in costs without at the same time increasing cycle time.

b) (3,2)-counter (fulladder); FA

$$c = a_1 a_2 \vee (a_1 \oplus a_2) \cdot a_3 \qquad s = (a_1 \oplus a_2) \oplus a_3 \qquad \text{(i)}$$

$$= a_1 a_2 \vee (a_1 \vee a_2) \cdot a_3 \qquad = (\bar{a}_1 a_2 \vee a_1 \bar{a}_2) \cdot \overline{\bar{a}_3 \vee (\bar{a}_1 a_2 \vee a_1 \bar{a}_2)} \cdot a_3$$

$$= a_1 a_2 \vee a_1 a_3 \vee a_2 a_3 \qquad = \bar{a}_1 \bar{a}_2 a_3 \vee \bar{a}_1 a_2 \bar{a}_3 \vee a_1 \bar{a}_2 \bar{a}_3 \vee a_1 a_2 a_3 \qquad \text{(ii)}.$$

Formulae (ii) yield the fulladder as disjunctive normal form; system (i) corresponds to two halfadders and one fulladder in series. We have:

$$\kappa_{FA} = \begin{cases} 14; \\ 25; \end{cases} \qquad \tau_{FA} = \begin{cases} 4 \text{ formulae (i)} \\ 2 \text{ formulae (ii)}. \end{cases}$$

a_3 is often a carry bit and its value is obtained after a_1 or a_2. In this case, the fresh carry c is calculated with (i) in 2 (additional) steps.

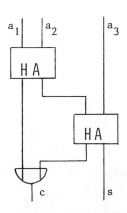

Fig. 2.3 (fulladder)

c) (4,2)-counter (input combination (1,1,1,1) excluded)

The simplest way of constructing a (4,2)-counter is from 3 halfadders (see fig. 2.4):

$$s = (a_1 \oplus a_2) \oplus (a_3 \oplus a_4);$$

$$c = a_1 a_2 \vee a_3 a_4 \vee (a_1 \oplus a_2) \cdot (a_3 \oplus a_4);$$

$$\kappa_{(4,2)} = 3 \cdot 6 + 5 = 23;$$

$$\tau_{(4,2)} = 4.$$

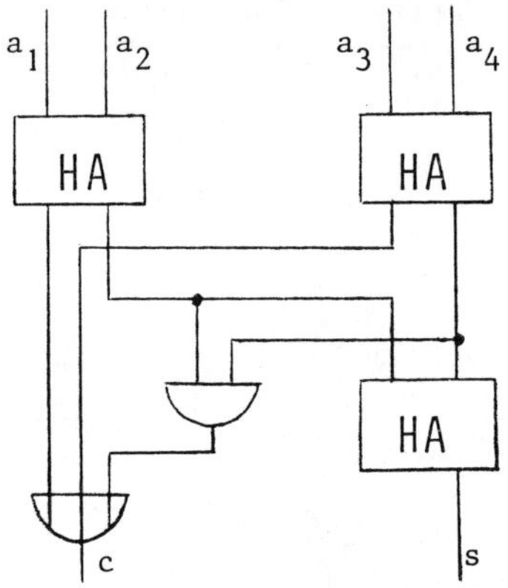

Fig. 2.4 (4,2)-counter

$(a_1,\ldots,a_4) \neq (1,\ldots,1)$

Specimens of (m,2)-counters (m = 2,3,4) using NAND-gates or threshold elements are illustrated in [Fe3].

2.2 Description of simple adder logic

2.2.1 Basic formulae, choice of notation

In this following section we shall deal with methods for the addition of two n-digit binary numbers a and b. Investigations from 1.3 show that, for this purpose, the formal sum s of both addends must be calculated; this applies to all binary digit codings so far treated (A+S, 2-complement, 1-complement). The additional operations (overflow test by doubling the sign, conversion to different code, end-around-carry using the 1-complement) will not be described further since they do not influence the actual methods of addition.

Subtraction can be explained by addition because of a-b = a+(-b);
this renders treatment of subtractors superfluous.

If $a = a_{n-1} \ldots a_0$ and $b = b_{n-1} \ldots b_0$, we have $s = s_n s_{n-1} \ldots s_0$
for the bits s_i of the formal sum (defined by the relationship):

$$\sum_{i=0}^{n-1} a_i \cdot 2^i + \sum_{i=0}^{n-1} b_i \cdot 2^i = \sum_{i=0}^{n} s_i \cdot 2^i \quad (s_i \in \{0,1\}):$$

$s_i = a_i \oplus b_i \oplus c_{i-1}$; $c_i = a_i \cdot b_i \vee (a_i \oplus b_i) \cdot c_{i-1}$ (i=0,.,n-1);

$s_n = c_{n-1}$; $c_{-1} = 0$.

c_i (the carry from position i to position i+1) and s_i are outputs of
a fulladder with inputs a_i, b_i and c_{i-1}.

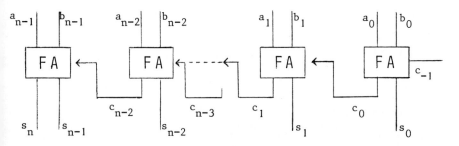

Fig. 2.5

The formulae show that the cycle time for addition is decisively
influenced by the time necessary for calculating the carries.

Notation

The most suitable coding for addition and subtraction of two binary
numbers is the 2-complement, since end around carry is unnecessary
and calculation of the total carry c_{n-1} is not required. A subtraction a-b is effected by inverting each bit of the respective addend b
and by adding a 1 to the least significant position of the representation.

Adding a 1 is performed without loss of time by suitably placing c_{-1} in advance. We write:

$$b_i^* := \begin{cases} b_i \\ \bar{b}_i \end{cases} \text{ and } c_{-1} := \begin{cases} 0 & \text{in the case of addition:} \quad a+b \\ 1 & \text{in the case of subtraction: } a-b. \end{cases}$$

In this chapter, all cycle time and cost calculations refer to a 2-complement representation of both addends.

2.2.2 Serial addition

The slowest and cheapest adder calculates (c_0,s_0), (c_1,s_1), ..., (c_{n-1},s_{n-1}) in sequence with one single fulladder (see fig. 2.6):

$$(c_i,s_i) = f_{(3,2)}(a_i,b_i,c_{i-1}) .$$

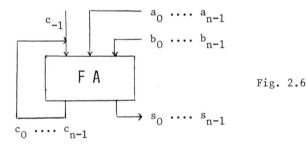

Fig. 2.6

The following microprogram contains a variable T which indicates whether subtraction (T=1) or addition (T=0) is to be performed.

Microprogram A1 (serial addition)

0 : $[a_{n-1},...,a_0]$:= addend 1; $[b_{n-1},...,b_0]$:= addend 2; Z := n;
1 : $b_i := b_i \oplus T$ (i=0,...,n-1); c := T;

2 : $a_i := a_{i+1}$ $\Big\}$ (i=0,..,n-2); $(c, a_{n-1}) := f_{(3,2)}(a_0, b_0, c)$; $Z := Z-1$;
   $b_i := b_{i+1}$ $b_{n-1} := 0$;

3 : <u>if</u> $Z > 0$ <u>then</u> <u>goto</u> 2;

4 : finished message (or test for overflow of the result).

The essential components of this program are represented in fig. 2.7 .
The counter Z provides that exactly n pulses are given to the
pulsed shift registers A and B as well as to the single digit
carry register c .

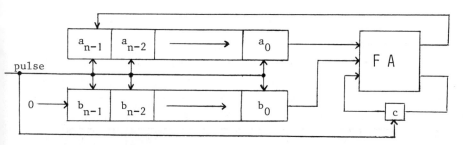

Fig. 2.7 (serial addition)

After n functions, $(a_{n-1}, \ldots, a_0)$ contains the (formal) sum (or difference)
of the two initial numbers. With this adder, the <u>expenditure</u> for the add
logic is <u>independent of the length n of the addends.</u> The <u>cycle time</u>,
however, increases <u>linearly with n.</u>

2.2.3 The von Neumann adder

Contrary to the serial adder, this oldest version of a parallel adder
contains its own halfadder (or fulladder, see 2.2.4) for each digit of
the two addends. Microprogram A2 describes the functioning of this
adder:

Microprogram A2 (von Neumann adder)

Doubling of signs is necessary for recognition of overflow situations, i.e. $a_{n-1} = a_{n-2}$, $b_{n-1} = b_{n-2}$;

0 : $A = [a_{n-1},\ldots,a_0] :=$ addend 1; $B = [b_{n-1},\ldots,b_0] :=$ addend 2;

1 : $b_i := b_i \oplus T$ $(i=0,\ldots,n-1)$;

2 : <u>if</u> $B \vee T \neq 0$

$$\text{then} \begin{bmatrix} a_i := a_i \oplus b_i & (i=0,\ldots,n-1) \\ b_{i+1} := a_i \cdot b_i & (i=0,\ldots,n-2) \\ b_0 := T;\ T := 0;\ \underline{goto}\ 2 & \end{bmatrix} \text{else} \begin{bmatrix} T := a_{n-1} \oplus a_{n-2}; \\ \underline{goto}\ 3 \end{bmatrix};$$

3 : <u>if</u> $T \neq 0$ <u>then goto</u> 5;

4 : finished message: no overflow;

5 : finished message: overflow;

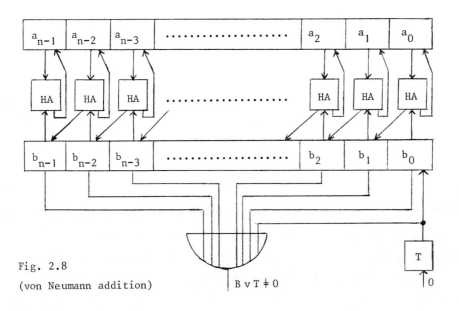

Fig. 2.8
(von Neumann addition)

Fig. 2.8 shows the essential components of microprogram A2. Function 2 of the program occurs at maximum (n+2) times (on the last pass we already have B=0). We have this most unfavourable case if both operands of a subtraction are 0. After function 2 has operated i times, we have $b_{i-1} = 1$; during pass (n+1), b_{n-1} is taken to the A-register, and not until the next attempt is the condition B=0 satisfied.

For the mean cycle time (number of functions) of A2 we have ([Cl1], [vN1]):

Proposition 2.1. *If positions a_i and b_i independently take 0 or 1, and if $P(a_i=0) = P(a_i=1) = \frac{1}{2}$ (similarly for b_i), we obtain for the mean cycle time of the von Neumann adder:*

$$\overline{\tau_{v.N}} = O(\log_2 n) \ .$$

2.2.4 Carry save addition, "Adder Tree"

Extension of the von Neumann adder by application of fulladders instead of halfadders is called carry save addition. This immediately allows the addition of 3 addends. Since the fulladders have 2 outputs only, but 3 inputs, a new addend can be treated with every pass of the loop (line 2 of microprogram A2). This form of addition (fig. 2.9) is especially suitable for the addition of numerous addends; it is also applied in multiplication and division procedures (see 3.4 and 4.5.4) .

For taking in the carry-in e_{-1} of the new addend $(e_{n-1},...,e_0)$, the free register position b_0 is used. For additions we have $e_{-1} = 0$, for subtractions $e_{-1} = 1$. From now on we will use the schematic representation shown.

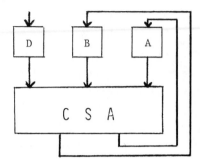

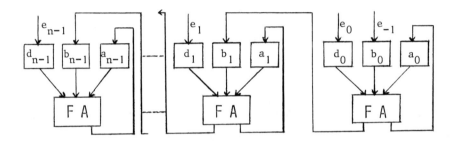

Fig. 2.9 (carry save addition)

The cycle time for the addition of m addends $S_1,\ldots,S_m$ can be reduced further by linearising the CSA machine: there are no feedbacks (re-storage into the A or B register); the resulting adder (fig. 2.10) contains m-2 "CSA steps", each comprising n fulladders. The CSA step outputs are connected to the inputs of the next step, together with the new addends.

For addition of the two remaining addends, a faster adder can be used, which will be dealt with in the next section (for example carry look ahead addition, see 2.3). Disregarding time and costs of the final addition, for m ($m \geq 3$) addends the following time and costs are obtained:

$\kappa = (m-2) \cdot n$ fulladder;

$$\tau = \begin{cases} 2 \cdot (m-2) & \text{according to form} \\ 4 \cdot (m-2) & \text{of the fulladder.} \end{cases}$$

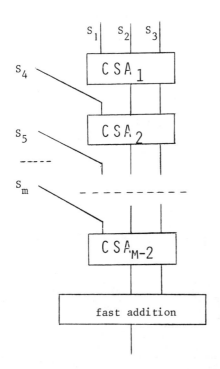

Fig. 2.10

From fig. 2.10 it is seen immediately that a further acceleration (by reducing the number of steps) can be achieved by having as many CSA as possible working parallel in one step. This yields a tree-type combination of CSA units ("Adder-Tree" [Wal]) which can be used to advantage with fast multipliers.

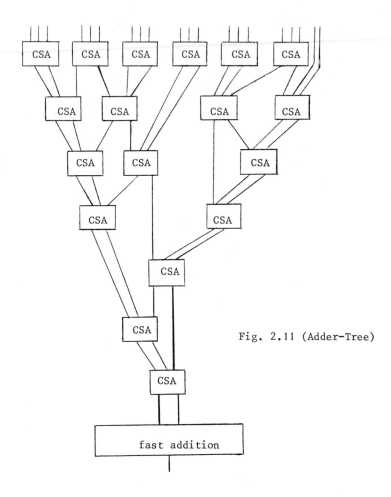

Fig. 2.11 (Adder-Tree)

Fig. 2.11 shows an "Adder-Tree" for m = 20 addends. With m - 2 = 18 carry save adders, the effort is just as great as for the "linear" process, but the number of steps is reduced from 18 to 7 . A detailed discussion regarding the "Adder-Tree" concept can be found under 3.4.4 .

2.2.5 Carry ripple addition (synchronous)

Like the carry save adder, the carry ripple adder consists of n fulladders. The carry is passed from one fulladder to the next without intermediate storage (fig. 2.12):

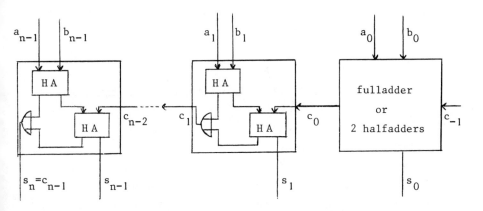

Fig. 2.12 (carry ripple addition)

Addition is performed in one single (relatively long) pulse. The most unfavourable case has to be considered when defining the pulse frequency. This occurs, when $c_{-1} = 1$ and $a_i \oplus b_i = 1$ ($i=0,\ldots,n-2$). In this case, carry c_{-1} is applied to all n digits. For transferring a carry, two logical steps are necessary in each case (see fig. 2.12). Calculation of c_0 is an exception; for constructing the first fulladder from 2 halfadders, 4 logical steps are necessary; however, when using the disjunctive normal form for c_0 and s_0, only 2 steps are necessary. This yields:

$$\tau_{C-R} = \begin{cases} 2n+2 \\ 2n \end{cases} \qquad \kappa_{C-R} = \begin{cases} 14n & \text{if} \quad \text{2 halfadders} \\ 14n+11 & \text{if} \quad \text{disj. normal form .} \end{cases}$$

2.2.6 Asynchronous carry ripple adder (carry completion)

With the carry ripple adder, final calculation of the sum bit s_i is possible only if the carry c_{i-1} cannot change further. This can be determined by a simple additional circuit to be constructed subsequently. Hence, the mean cycle time or carry ripple adder is shortened to approximately $2 \times \log_2 n$ steps (see corresponding result for the von Neumann addition).

Definition 2.2. Let $a = a_{n-1} \ldots a_0$, $b = b_{n-1} \ldots b_0$:

$(a_{j+k}, b_{j+k}), \ldots, (a_{j+1}, b_{j+1})$ means *propagation chain of length k*

$$\Leftrightarrow \begin{cases} a_{j+r} \oplus b_{j+r} = 1 & \text{for } r = 1, \ldots, k \\ a_{j+r} \oplus b_{j+r} = 0 & \text{for } r = 0, r = k+1 \end{cases}$$

A measure of the cycle time of the adder is the length of the longest propagation chain. By means of the two variables f_i and h_i we can observe for each bit pair (a_i, b_i) whether it belongs to a propagation chain and whether the accompanying carry c_{i-1} has already been determined.

Definition 2.3. $f_i := a_i \cdot b_i \vee (a_i \oplus b_i) \cdot f_{i-1} = a_i \cdot b_i \vee (a_i \vee b_i) \cdot f_{i-1}$;

$h_i := \overline{a_i} \cdot \overline{b_i} \vee (a_i \oplus b_i) \cdot h_{i-1} = \overline{a_i \vee b_i} \vee \overline{a_i \cdot b_i} \cdot h_{i-1}$.

At the beginning of the addition, we have:

$$f_i = h_i = 0 \quad (i \geq 0), \quad f_{-1} = c_{-1}, \quad h_{-1} = \overline{c_{-1}}.$$

Conclusion. *If $f_i \vee h_i \neq 0$, determination of the carry for the bit pair (a_i, b_i) is completed and we have: $f_i = c_i$; $h_i = \overline{c_i}$. The addition can be completed if we have:* $\prod_{i=0}^{n-2} (f_i \vee h_i) = 1$.

Calculation of (f_i, h_i) from (f_{i-1}, h_{i-1}) results in the costs
$\kappa = 12$.

Cycle time and costs:

Considering that parts of the additional logic are already included in the carry ripple adder, we can readily conclude:

$$\kappa_{ASYN} = 14n + 11(n-1) = 25n - 11, \qquad \tau_{ASYN} \leq 2 \cdot L + 4 .$$

L denotes the length of the longest propagation chain.

2.2.7 Exclusive OR adder ([Ki1], [Ki2], [Sa1], [Le2], [Fe2])

For hardware, besides exclusive OR gates, high speed switches are used in this adder which can be opened or closed from an external signal.

$$\text{switch} \begin{cases} \text{closed when } U = 1 \\ \text{open when } U = 0 \end{cases}$$

The adder logic can be described as follows: according to 2.2.1 we have:

$$c_i = a_i \cdot b_i \vee (a_i \oplus b_i) \cdot c_{i-1} ;$$
$$s_i = (a_i \oplus b_i) \oplus c_{i-1} ;$$

i.e.
$$c_i = \begin{cases} 1 & \text{when } a_i \cdot b_i = 1 \\ 0 & \text{when } \overline{a_i} \cdot \overline{b_i} = 1 \\ c_{i-1} & \text{when } a_i \oplus b_i = 1 . \end{cases}$$

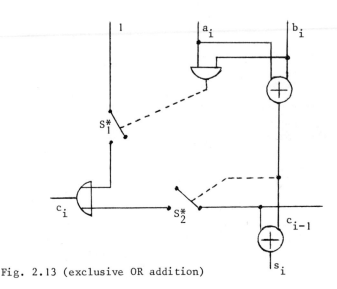

Fig. 2.13 (exclusive OR addition)

Fig. 2.13 demonstrates calculation of c_i and s_i. Switch S_1^* is closed when $a_i \cdot b_i = 1$, i.e. $c_i = 1$; the same applies to switch S_2^*. Kilburn [Kil] uses a third switch by which c_i is set at 0, when a_i and b_i are 0 simultaneously.

Cycle time and cost analysis

The exclusive OR adder can scarcely be compared with other adders, since it contains different modules.

In the most unfavourable case (see 2.2.5), the carry must pass through all switches and all OR gates. In addition there is the time for opening and closing the switches.

The adder is of special advantage if switches are available with a very low switching and cycle time. Assuming that the delay in passing through a switch corresponds approximately to the cycle time of an AND or OR gate, we have an addition time corresponding to the carry ripple adder. With regard to the costs, the two adders can be compared in a limited fashion, when the availability and price of the modules used should be considered.

Remark: With regard to the adders treated below, high speed switches can be used also instead of AND or OR gates for calculating or passing carries; we shall not deal further with this.

2.3 Carry look ahead addition

Carry look ahead addition reduces the addition cycle time by accelerating calculation of the carries; this results in partial or complete resolution of the recursive definition of c_j. At present, this concept is the most common method for high speed addition of two binary numbers. We will first demonstrate the concept on an unrealistic extreme case, from which carry look ahead formulae can be readily derived.

2.3.1 Carry look ahead on all n digits (n bit carry look ahead)

<u>Definition 2.4.</u> $k_i := a_i \cdot b_i$, $d_i := a_i \vee b_i$, $e_i := a_i \oplus b_i = \overline{k_i \vee \bar{d}_i}$.

For the sum or carry bits, we obtain from 2.2.1:

$$s_i = a_i \oplus b_i \oplus c_{i-1} = e_i \cdot \overline{c_{i-1}} \vee \bar{e}_i \cdot c_{i-1} \quad (i = 0,\ldots,n-1) ;$$

$$c_j = a_j \cdot b_j \vee (a_j \oplus b_j) \cdot c_{j-1} = k_j \vee e_j \cdot c_{j-1}$$

$$ = a_j \cdot b_j \vee (a_j \vee b_j) \cdot c_{j-1} = k_j \vee d_j \cdot c_{j-1} .$$

$$\bar{c}_j = \bar{a}_j \cdot \bar{b}_j \vee \overline{(a_j \oplus b_j)} \cdot \bar{c}_{j-1} = \bar{d}_j \vee \bar{e}_j \cdot \bar{c}_{j-1}$$

$$\phantom{\bar{c}_j} = \overline{a_j \vee b_j} \vee \overline{a_j \cdot b_j} \cdot \bar{c}_{j-1} = \bar{d}_j \vee \bar{k}_j \cdot \bar{c}_{j-1} .$$

Resolving the recursion formulae for c_j and $\bar{c}_j$, we have:

$$c_j = k_j \vee d_j k_{j-1} \vee d_j d_{j-1} k_{j-2} \vee \ldots \vee d_j d_{j-1} \ldots d_1 k_0 \vee d_j \ldots d_0 c_{-1} \quad (i) ;$$

$$\bar{c}_j = \bar{d}_j \vee \bar{k}_j \bar{d}_{j-1} \vee \bar{k}_j \bar{k}_{j-1} \bar{d}_{j-2} \vee \ldots \vee \bar{k}_j \bar{k}_{j-1} \ldots \bar{k}_1 \bar{d}_0 \vee \bar{k}_j \ldots \bar{k}_0 \bar{c}_{-1} \quad (ii) .$$

Variables d_i in (i) and variables $\bar{k}_i$ in (ii) can be replaced by the exclusive OR gates e_i.

Dispensing with auxiliary variables k_i and d_i, we can immediately manage with two logical steps only, but we will then have to tolerate an effort, increasing exponentially with j, for c_j and $\bar{c}_j$. Hence we exclude this possibility.

Cycle time and costs

Gates with n entries are not available for large n. At least, however, costs for gates no longer increase linearly with the number of inputs, as was assumed in the cost function. Furthermore, the gate cycle time is no longer independent of the number of inputs. Hence, the cycle time and costs formulae quoted below will be considered only as a guide. Effort and cycle time of the complete carry look ahead addition depend on whether s_i is calculated from e_i and c_{i-1} in two logical steps (method A), or whether s_i is determined by applying the formulae (i) or (ii) for c_{i-1} or $\bar{c}_{i-1}$ in only one logical step (method B).

Time schedule: method A   method B

Step	Calculation of	Step	Calculation of
1	k_i, d_i	1	k_i, d_i
2	e_i	2	e_i
3	$\left[c_i = f_1(k_i, d_i), \bar{c}_i = f_2(k_i, d_i)\right]$	3	$\left[s_i = f_4(k_i, d_i, e_i)\right]$
4	$\left[s_i = f_3(e_i, c_{i-1})\right]$	4	
5			

Proposition 2.2.   1. *Method A:*

$$\tau_A = 5 \quad , \quad \kappa_A = \frac{1}{6} n^3 + n^2 + 11\frac{5}{6} n - 1 \; ;$$

2. Method B:

$$\tau_B = 4 \quad , \quad \kappa_B = \frac{1}{3} n^3 + 3 n^2 + 8\frac{2}{3} n .$$

Proof is elementary. It is evident that almost the entire costs are due to calculation of the carries. The outlay for method B is more than double that for method A since, in addition to the carries c_i, the complements $\bar{c}_i$ have to be calculated also.

2.3.2 Carry look ahead of 1st order (group size g)

Technical restrictions and considerations with regard to costs prohibit the application of the n bit carry look ahead addition for longer addends. Hence, the addends are divided into smaller groups of the size g with g (often g=4,5 or 6) chosen in such a way that the carry look ahead principle can be applied from the technical standpoint and that the outlay is acceptable. We assume without marked restriction that n can be divided by g (otherwise the logic for the group on the far left is simplified). Hence we have:

$$n = n_1 \cdot g \quad (n_1 \in \mathbb{N}) .$$

The total carry of a group is passed through the groups according to the carry ripple principle. Hence, a 1st order binary carry look ahead adder is a carry ripple adder with base $d = 2^g$.

For brevity, we define as follows:

Definition 2.5. $A_i := (a_{(i+1)g-1}, \ldots, a_{ig})$ $\quad C_i := c_{(i+1)g-1}$
$\qquad\qquad B_i := (b_{(i+1)g-1}, \ldots, b_{ig})$ $\quad (i=-1,\ldots,n_1-2)$.
$\qquad\qquad S_i := (s_{(i+1)g-1}, \ldots, s_{ig})$ $\quad C_i$ is the total carry of
$\qquad\qquad (i = 0, \ldots, n_1-1)$ $\qquad\qquad$ group i to group $i+1$.

The 1st order carry look ahead adder can be demonstrated schematically as follows:

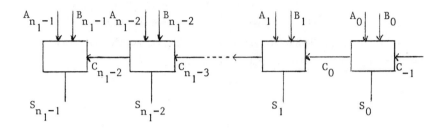

Fig. 2.14 (1st order carry look ahead addition)

$$\hat{S}_i := A_i \mathbin{\hat{\oplus}} B_i \mathbin{\hat{\oplus}} C_{i-1} \quad (\text{"} \mathbin{\hat{\oplus}} \text{"} := \text{addition modulo } 2^g).$$

The formulae for the calculation of C_j are derived from 2.3.1:

$$C_j = c_{(j+1)g-1} = k_{(j+1)g-1} \vee d_{(j+1)g-1} \cdot k_{(j+1)g-2} \vee \ldots \vee$$

$$\vee d_{(j+1)g-1} \cdot \ldots \cdot d_{jg+1} \cdot k_{jg} \vee d_{(j+1)g-1} \cdot \ldots \cdot d_{jg} \cdot C_{j-1}$$

(d_i can be replaced by variables e_i as in 2.3.1).

Table 2.1 shows that for all groups except the last, the accompanying sum bits are calculated to a maximum of $2n_1+1$ steps, if in addition to the total carry all individual carries are accelerated also. For group $0,\ldots,n_1-2$, a less expensive (and slower) calculation of the single carries can be performed without increasing the total cycle time of the addition; however, we will not go into details regarding these associated variables.

Table 2.1: time schedule for 1st order carry look ahead addition

Step	Group (n_1-1)	Group (n_1-2)	....	Group 1	Group 0
1	k_i, d_i	k_i, d_i		k_i, d_i	
2	e_i	e_i		e_i	carries
3					group 0
4				carries	sum bits
5				group 1	group 0
6				sum bits	
7				group 1	
.			....		
.		carries			
.		group (n_1-2)			
$2n_1$	carries	sum bits			
$2n_1+1$	group (n_1-1)	group (n_1-2)			
$2n_1+2$	sum bits				
$2n_1+3$	group (n_1-1)				

The cycle time and costs results set out below are based on table 2.1. Note that as with other adders, the last group carry C_{n-1} need not be calculated here either, since we assumed the 2-complement as notation with doubling of the sign.

Cycle time and costs:

$$\tau_{CLA1} = 2 \cdot n_1 + 3 = 2 \cdot \frac{n}{g} + 3 \ ;$$

$$\kappa_{CLA1} \leq \frac{1}{6}(n-3)g^2 + \frac{1}{2}(3n-5)g + \frac{1}{3}(43n-3) \ .$$

As previously, costs are governed essentially by the outlay necessary for the carries.

2.3.3 2nd order carry look ahead

The concept discussed in the previous section can be iterated by combining g' groups with a "section" accelerating total carries of the sections and passing through the sections by means of a carry ripple adder with base $d = (2^g)^{g'}$. In applying this concept, it is advisable to introduce a range of new auxiliary functions and designations. We assume $n = n_1 \cdot g = n_2 \cdot g' \cdot g$ ($n_2 \in \mathbb{N}$).

Definition 2.6. *We define the following new auxiliary variables:*

a. $A_i^{(2)} := (A_{(i+1)g'-1}, \ldots, A_{ig'})$; $C_i^{(2)} := C_{(i+1)g'-1}$

   $B_i^{(2)} := (B_{(i+1)g'-1}, \ldots, B_{ig'})$; $= c_{(i+1)g' \cdot g-1}$

   $S_i^{(2)} := (S_{(i+1)g'-1}, \ldots, S_{ig'})$; $(i=-1, \ldots, n_2-2)$;

   $(i = 0, \ldots, n_2-1)$; $C_i^{(2)}$ *is the total carry of section i.*

b. $D_j := d_{(j+1)g-1} \cdot \ldots \cdot d_{jg}$;

   $K_j := k_{(j+1)g-1} \vee d_{(j+1)g-1} \cdot k_{(j+1)g-2} \vee \ldots \vee d_{(j+1)g-1} \cdot \ldots$

$$\cdot d_{jg+1} \cdot k_{jg}$$

   $(j = 0, \ldots, n_1-1)$.

   $S_i^{(2)} := A_i^{(2)} \,\hat{\oplus}\, B_i^{(2)} \,\hat{\oplus}\, C_{i-1}^{(2)}$ (" $\hat{\oplus}$ " = addition modulo $(2^g)^{g'}$) .

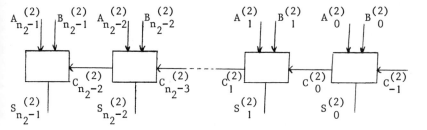

Fig. 2.15 (2nd order carry look ahead addition)

The group carries of section i are calculated with the new auxiliary functions as follows:

$$C_{(i+1)g'-r} = K_{(i+1)g'-r} \vee D_{(i+1)g'-r} \times K_{(i+1)g'-r} \vee \ldots \vee$$

$$\vee D_{(i+1)g'-r} \cdot \ldots \cdot D_{ig'+1} \cdot K_{ig'}$$

$$\vee D_{(i+1)g'-r} \cdot \ldots \cdot D_{ig'+1} \cdot D_{ig'} \cdot C_{(i-1)}^{(2)}$$

$$(r=1,\ldots,g')$$

It is evident also that the total carry

$$C_i^{(2)} = C_{(i+1)g'-1}$$

of a section can be calculated in two logical steps from $C_{i-1}^{(2)}$.

The costs and the cycle time of a 2nd order CLA adder can be calculated on the basis of a time schedule (table 2.2). The costs for large n increase only slightly as compared with those for the 1st order carry look ahead. As regards additional costs, there is only the expenditure for accelerated calculation of the group carries within the sections.

The costs for new auxiliary functions are compensated by simplified calculation of the group carries (with auxiliary functions K_i and D_i). We thus have:

$$\tau_{CLA2} = 2n_2 + 7 = 2 \cdot \frac{n}{g \cdot g'} + 7 \ ;$$

$$\kappa_{CLA2} = \kappa_{CLA1} + \frac{1}{2}(g^2 - g'^2) + \frac{5}{2}(g-g') + \frac{n}{6g}(g'^2 + 9g' + 2) \ .$$

In the most important special case, this is simplified as follows:

$$\kappa_{CLA2} = \kappa_{CLA1} + \frac{1}{6} n \cdot (g + 9 + \frac{2}{g}) \ .$$

The additional concept dealt with can be extended to carry look ahead methods higher than 2nd order (combination of g" sections each, definition of new auxiliary functions, accelerated calculation of the total carry, the new, larger sections etc.). In practice, these methods are less significant, since the word length of the addends is restricted ($n \leq 200$, generally even $n \leq 60$), and the advantages of higher-than-2nd-order carry look ahead methods apply only with larger values of n. For relatively small values of n, a hierarchy of 3 or more types of groups is uneconomical.

2.4 Carry skip (simplified carry look ahead addition)

The carry skip technique is a compromise between carry ripple and carry look ahead addition. The carries of a group with size g are calculated in 2g steps at most as with the carry ripple adder. A simple, additional circuit is sufficient for the accelerated passage of the carries over the groups. If the adder is suitably divided into groups, a considerable reduction in the maximum cycle time is achieved as compared with the carry ripple adder.

Table 2.2: Time schedule for 2nd order CLA

Step	section n_2-1	section n_2-2	.....	section 1	section 0
1	k_i, d_i	k_i, d_i		k_i, d_i	
2	e_i, D_i	e_i, D_i		e_i, D_i	
3	K_i	K_i		K_i	
4					group
5					carries
6				group	single
7				carries	carries
8				single	sum
9				carries	bits
⋮			.....	sum bits	
$2n_2$		group			
$2n_2+1$		carries			
$2n_2+2$	group	single			
$2n_2+3$	carries	carries			
$2n_2+4$	single	sum			
$2n_2+5$	carries	bits			
$2n_2+6$	sum				
$2n_2+7$	bits				

2.4.1 Constant group size g

We will apply the auxiliary functions introduced in 2.3.3:

$$D_j := d_{(j+1)g-1} \cdot \ldots \cdot d_{jg} \;;$$

$$K_j := k_{(j+1)g-1} \vee d_{(j+1)g-1} \cdot k_{(j+1)g-2} \vee \ldots \vee d_{(j+1)g-1} \cdot \ldots \cdot d_{jg+1} \cdot k_{jg}$$

We thus have for the total carry C_j of group j:

$$C_j = k_{(j+1)g-1} \vee d_{(j+1)g-1} \cdot k_{(j+1)g-2} \vee .. \vee d_{(j+1)g-1} \cdot ... \cdot d_{jg+1} \cdot k_{jg}$$

$$\vee \; d_{(j+1)g-1} \cdot ... \cdot d_{jg} \cdot C_{j-1}$$

$$= K_j \vee D_j \cdot C_{j-1} \; .$$

The carry skip adder calculates a carry $C_j = 1$ as follows:

a. $K_j=1$, $D_j \cdot C_{j-1}=0$; i.e. a new carry is generated in group j: $C_j = K_j$ is calculated by a carry ripple adder of length g in $2 \cdot g$ logical steps at most.

b. $D_j \cdot C_{j-1}=1$; i.e. C_{j-1} is passed over group j: in this case, passage of the carry is performed by the auxiliary function D_j (already calculated) in two logical steps.

The structure of the carry skip group is shown in fig. 2.16. The OR gates used for the calculation of D_j are already included in the associated fulladders (see 2.1). Hence they are disregarded when calculating the costs.

<u>Cycle time and costs</u>

When calculating the cycle time the most unfavourable case is considered, as usual. This occurs for addends $a = (a_{n-1},...,a_0)$ and $b = (b_{n-1},...,b_0)$ of the form:

$$a_0 = b_0 = 1, \quad c_{-1} = 0, \quad a_i \oplus b_i = 1 \quad (i = 1,...,n-2) \; .$$

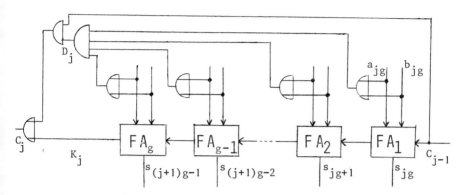

Fig. 2.16 (1st order carry skip addition)

In this case there is a carry only in group 0 ($K_0 = 1$), and it is not passed over this group directly ($D_0 \cdot c_{-1} = 0$), but over all the other groups. Calculation of K_0 and determination of the sum bits of the last group require maximum cycle time. All other sum and carry bits occur before s_{n-1}. With $n_1 := n/g$, we have:

$$\tau_{C\text{-SKIP, max}} = \underbrace{2 \cdot g + 1}_{(*)} + \underbrace{2 \cdot (n_1 - 2)}_{(**)} + \underbrace{2 \cdot g}_{(***)} = 4g + 2n_1 - 3 .$$

(*)   : Calculation of K_0 in 2 x g steps and $C_0 = K_0 \vee D_0 \cdot c_{-1}$ in an additional step.

(**) : Passage of carry C_0 over groups $1, \ldots, n_1 - 2$.

(***) : Calculation of s_{n-1} in 2g steps (two each per fulladder).

The minimum cycle time is obtained for group size $g_{opt} = \sqrt{\frac{n}{2}}$, i.e. for $n = 2 \cdot g_{opt}^2$, $n_1 = 2 \cdot g_{opt}$:

$$\tau_{C\text{-SKIP, max}}(g_{opt}) = 4 \cdot \sqrt{2} \cdot \sqrt{n} - 3 \approx 5.656 \cdot \sqrt{n} - 3 .$$

Examples: a. $n = 32 \Rightarrow g_{opt} = 4$, $\tau = 29$;

b. $n = 50 \Rightarrow g_{opt} = 5$, $\tau = 37$.

The additional costs for the group j carry skip logic, as compared with the carry ripple adder, are restricted to the AND gate with g inputs for the calculation of D_j and the two gates with two inputs for determining $C_j = K_j \vee D_j \cdot C_{j-1}$.

$$\kappa_{C\ SKIP} = \kappa_{C-R} + n_1 \cdot (g+4) = 14n + \frac{n}{g} \cdot (g+4) = 15n + \frac{4n}{g}.$$

2.4.2 Variable group size

Discussion of the most unfavourable case under 2.4.1 shows that the carry passes very quickly over the inner groups, while the two marginal groups in which the carry is generated or terminates require a relatively long time. In the following we will examine the extent to which addition is accelerated by reducing the marginal groups and at the same time extending the inner groups ([Le1], [Ma2]).

Definition 2.7. *Let k_j be the length of group G_j of a C skip adder.*

$$K_n := (k_{n_1-1},\ldots,k_0)$$

is a group scheme, when:

$k_j \geq 1$, $n_1 \geq 2$ and $k_0 + k_1 + \ldots + k_{n_1-1} = n$.

Let $S(K_n)$ denote the maximum cycle time of the carry skip addition applying scheme K_n.
A group scheme with minimum cycle time is optimum.

In 2.4.1, we demonstrated $S((g,\ldots,g)) = 4g + 2n_1 - 3$. This result can be generalized to variable group sizes. In the following we assume that at least one $k_i > 1$.

Lemma 2.3. $S(K_n) = \max\limits_{0 \le \alpha < \beta \le n_1-1} [2 \cdot v_{(\alpha,\beta)} + sign(k_\alpha -1)]$,

where $v_{(\alpha,\beta)} := k_\alpha + k_\beta + \beta - \alpha - 1$.

Proof. If a carry is generated at G_α and runs to group G_β ($\beta \bowtie \alpha$) inclusive, the sum bits of the section consisting of $G_\beta, \ldots, G_\alpha$ are calculated only after

$$h = \begin{cases} 2 \cdot k_\alpha + 1 + 2 \cdot (\beta+\alpha-1) + 2k_\beta & \text{where } k_\alpha > 1 \\ 2 + 2 \times (\beta-\alpha-1) + 2k_\beta & \text{where } k_\alpha = 1 \end{cases}$$

logical steps in the most unfavourable case. (Where $k_\alpha = 1$ and $\alpha = 0$, we assume that the first fulladder with disjunctive normal form is obtained in two logical steps.)

For determining an optimum scheme K_n, the following <u>optimizing problem (0*)</u> is to be solved:

0^*: Determine $n_1, k^*_{n_1-1}, \ldots, k^*_0$ with

$$S((k^*_{n_1-1}, \ldots, k^*_0)) = \min_{(k_{n_1-1}, \ldots, k_0)} \max_{0 \le \alpha < \beta \le n_1-1} [2 \cdot v_{(\alpha,\beta)} + sign(k_\alpha-1)]$$

$$= \min_{(k_{n_1-1}, \ldots, k_0)} \max_{0 \le \alpha < \beta \le n_1-1} 2 \cdot v_{(\alpha,\beta)} + 1$$

$n_1 \ge 2;\ \Sigma k_i = n$

$$= 2 \cdot T(K^*_n) + 1,$$

where $T(K^*_n) := \min\limits_{(k_{n_1-1}, \ldots, k_0)} \max\limits_{0 \le \alpha < \beta \le n_1-1} v_{(\alpha,\beta)}$.

Hence, we are seeking the group number n_1 and group sizes k_i^*, so that the maximum cycle time $S(K_n^*)$ of this scheme is minimum. We need two auxiliary propositions for resolving this problem:

Lemma 2.4. *There is one solution K_n^* to the optimizing problem (0*) in which adjacent group sizes differ by 1 at most.*

Proof. Let $K_n = (k_{n_1-1},\ldots,k_0)$ be a group scheme. We assume that there is a h $(0 \le h \le n_1-2)$, for which $|k_{h+1}-k_h| \ge 2$. Now we define a new scheme K_n' by:

$$K_n' = \begin{cases} (1, n-2, 1) & \text{when } K_n = (n-1, 1) \\ (\ldots,k_{h+1}-1, k_h+1,\ldots) & \text{when } K_n \ne (n-1, 1), k_{h+1} > k_h \\ (\ldots,k_{h+1}+1, k_h-1,\ldots) & \text{when } K_n \ne (n-1, 1), k_{h+1} < k_h \end{cases}$$

Simple examination shows that:

$$S(K_n') \le S(K_n) ;$$

Repeated application of this principle finally yields a scheme in which adjacent group sizes differ by 1 at most and the cycle time does not exceed that of K_n.

Hence, we can restrict ourselves to schemes for which adjacent groups differ by 1 at most. Lemma 2.5 shows that in this case, $\max v_{(\alpha,\beta)}$ is assumed for $\alpha = 0$, $\beta = n_1-1$:

Lemma 2.5. *If $|k_i - k_{i+1}| \le 1$ $(i = 0,\ldots,n_1-2)$, we have:*

$$\max_{0 \le \alpha < \beta \le n_1-1} v_{(\alpha,\beta)} = v_{(0, n_1-1)} = k_0 + k_{n_1-1} + n_1 - 2 .$$

Proof. Assume $\max v_{(\alpha,\beta)}$ is obtained for $\beta = \beta^*$.

$\Rightarrow v_{(\alpha-1, \beta^*)} = k_{\alpha-1} + 1 + k_{\beta^*} + \beta^* - \alpha - 1 \ge v_{(\alpha,\beta^*)}$ (since $k_{\alpha-1}+1 \ge k_\alpha$) ;

$\Rightarrow \max v_{(\alpha,\beta^*)} = v_{(0,\beta)} = k_0 + k_{\beta^*} + \beta^* - 1$.

It is seen by analogy that $\beta^* = n_1-1$.

We obtain immediately from lemma 2.5:

Lemma 2.6. *a.* $T(K_n^*) = \min\limits_{(k_{n_1-1},\ldots,k_0)} (k_0+k_{n_1-1}+n_1-2)$.

$n_1 \geq 2$, $\Sigma k_i = n$, $|k_i-k_{i+1}| \leq 1$

b. An optimum value for $T(K_n^*)$ can be obtained by making the marginal groups as small as possible and allowing the group size to increase by exactly 1 each from the margins to the middle. Hence, n_1 also is as small as possible.

We are now able to construct optimum schemes and indicate their cycle times:

Proposition 2.7. *a. Let $n = m(m+1)$ ($m \in N$). For the optimum scheme K_n^*, we have:*

$K_n^* = (k_{n_1-1}^*,\ldots,k_0^*) = (1,2,\ldots,m-1,m,m,m-1,\ldots,1)$, *i.e.* $n_1 = 2m$.

$\tau_n = S(K_n^*) = 4m+1$.

b. Let $n = m^2$ ($m \in N$).

$K_n^* = (k_{n_1-1}^*,\ldots,k_0^*) = (1,2,\ldots,m-1,m,m-1,\ldots,1)$; $n_1 = 2m-1$.

$\tau_n = S(K_n^*) = 4m-1$.

c. Where $m^2 < n < m \cdot (m+1)$ or $m \cdot (m-1) < n < m^2$, we have for the cycle time:

$\tau_n = 4m+1$ *or* $\tau_n = 4m-1$.

An optimum scheme can be obtained from the larger scheme (for $m(m+1)$ or m^2) by deleting groups from the margins, beginning with the smallest group. Deletion is continued so long as the sum of the lengths of the deleted groups does not exceed $m(m+1)-n$ or m^2-n. When no further *group can be completely removed, we reduce the total length of the remaining groups to the value n by "rounding down" the groups of maximum size (observing the condition $|k_i - k_{i+1}| \leq 1$).*

Proof. a,b: The optimum schemes are defined uniquely.

$$\tau_n = S(K_n^*) = 2 \cdot \min_{\substack{K_n \\ n_1 \geq 2; \; \Sigma k_i = n}} \max_{0 \leq \alpha < \beta \leq n_1 - 1} v_{(\alpha,\beta)} + 1 = 2 \cdot v_{(0, n_1 - 1)} + 1$$

$$= 2 \cdot \min [k_0 + k_{n_1 - 1} + n_1 - 2] + 1 = 2 \cdot [1 + 1 + n_1 - 2] + 1 = 2 n_1 + 1.$$

c. Deletion of a marginal group or rounding down an inner group does not change $S(K_n^*)$ (reduction of n_1 is compensated by extension of the new marginal group).

Examples: 1. $n = 56 = 7 \cdot 8 \Rightarrow m = 7$, $n_1 = 2 \cdot m = 14$;

$$\tau_{56} = 4 \cdot 7 + 1 = 29.$$

Optimum scheme: $(1,2,3,4,5,6,7,7,6,5,4,3,2,1)$.

Remark: For example, a carry which is generated in the second group from the right and passed over all other groups up to the last one has maximum cycle time; $\tau_n = 2 \cdot 2 + 1 + 2 \cdot 11 + 2 = 29$.

2. $n = 50 \in [49, 56]$; $\tau_{50} = \tau_{56} = 4 \cdot 7 + 1 = 29$.

Optimum schemes:

a. $(\mathbf{1,2},3,4,5,6,7,7,6,5,4,3,\mathbf{2,1})$, $n_1 = 10$, $\tau_{50} = 2 \cdot (3+3+10-2) + 1 = 29$;

b. $(\mathbf{1,2,3},4,5,6,7,7,6,5,4,3,2,1)$, $n_1 = 11$, $\tau_{50} = 2 \cdot (1+4+11-2) + 1 = 29$;

c. $(1,2,3,4,5,6,7,7,6,5,4,\mathbf{3,2,1})$, $n_1 = 11$, $\tau_{50} = 2 \cdot (4+1+11-2) + 1 = 29$.

3. $n = 44 \in [42, 49]$, $\tau_{44} = \tau_{49} = 4 \cdot 7 - 1 = 27$.

Optimum schemes, for example, are:

a. $(1,2,3,4,5,6,7,6,5,4,3,2,1)$, $n_1=10$, $\tau_{44} = 2 \cdot (3+2+10-2)+1 = 27$;
                      6

b. $(1,2,3,4,5,6,7,6,5,4,3,2,1)$, $n_1=11$, $\tau_{44} = 2 \cdot (1+3+11-2)+1 = 27$.
                     6 5

The statements of a. – c. of proposition 2.7 can be combined to a general cycle time result:

Proposition 2.8. *For maximum carry skip adder cycle time with variable group size, we have:*

$$\tau_n = \begin{cases} 2 \cdot \lceil \sqrt{4n+1} \rceil -1, & \text{where } m^2 < n \leq m \cdot (m+1) \\ 2 \cdot \lceil \sqrt{4n} \rceil -1, & \text{where } (m-1)m < n \leq m^2 \end{cases} \quad i.e. \quad \tau_n \approx 4 \cdot \sqrt{n},$$

(A condition for this is that AND or OR gates with any number of inputs are available and that the cycle time of the gates does not depend on the number of inputs.)

Proof. $m^2 < n \leq m \cdot (m+1) \Leftrightarrow \lceil \sqrt{4n+1} \rceil = 2m+1$;

$(m-1)m < n \leq m^2 \Leftrightarrow \lceil \sqrt{4n} \rceil = 2m$.

The first case yields: $\tau_n = 4m+1 = 2 \cdot (2m+1) - 1 = 2 \cdot \lceil \sqrt{4n+1} \rceil -1$,

The second case yields: $\tau_n = 4m-1 = 2 \cdot 2m-1 = 2 \cdot \lceil \sqrt{4n} \rceil -1$.

Cycle time and cost analysis

Comparison with cycle time

$$\tau_{C \text{ SKIP, max}}(g_{opt}) = 4 \cdot \sqrt{2} \cdot \sqrt{n} - 3$$

of the carry skip adder with constant group size $g = g_{opt} \approx \sqrt{n/2}$ shows that the variable group sizes accelerate the addition time by the factor $\sqrt{2}$ (disregarding cycle time problems and availability of gates).

With the function for costs chosen by us, which takes the total number of gate inputs as a measure of the costs, the costs for variable group sizes exceed the corresponding costs with fixed group sizes only minimally. For $k_0 > 1$, we have:

$$\kappa = \kappa_{C-R} + \sum_{i=0}^{n_1-1} k_i + 4 \cdot n_1 = \kappa_{C-R} + n + 4 \cdot n_1 < \kappa_{C-R} + 5n \ .$$

2.4.3 Higher order carry skip addition

The carry skip addition principle can be iterated (combining groups to sections, see 2.3.3); combination is also possible with carry look ahead techniques. Figures 2.17 and 2.18 demonstrate two examples; further variants are possible.

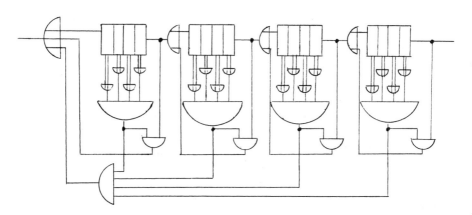

Fig. 2.17 (2nd order carry skip addition)

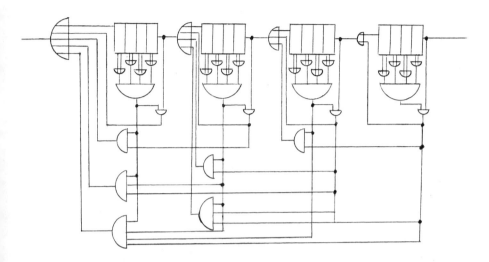

Fig. 2.18 (Combination of carry skip and carry look ahead)

2.5 Conditional sum addition [Sk1]

The conditional sum adder operates according to a different principle from the adders so far treated, which reduce the maximum carry cycle time and hence the maximum length of the addition by fairly costly carry calculations.

The adder is divided into groups G_i ($i=0,\ldots,n_1-1$). For each of these groups G_i, the sum bits and the total carry c_i are calculated for <u>both alternatives</u> ($c_{i-1}=0$ or $c_{i-1}=1$). The adder starts (**step** 0) with groups of size 1. During transition from step i to step $i+1$, the size of the group is doubled by combining two adjacent groups into a fresh one. Calculation of the sum bits or total carry of a new group is performed by simple selection from the sum bits of the associated lower step group. Since the group size in step i has the value 2^i, the adder requires $\lceil \log_2 n \rceil + 1$ logical steps $(0, 1, \ldots, \lceil \log_2 n \rceil)$ for calculating the sum bits for the two alternatives $c_{-1}=0$ or $c_{-1}=1$. Finally, the two correct sum bits are selected as a function of c_{-1} by a gate circuit.

Writing $\boxed{c_{i,j}^m \mid S_{i,j}^m}$ for the carry bit and the 2^m sum bits of group i in step m, and if the initial carry for this group has the value j $(j \in \{0,1\})$, the adder operates as follows:

$c_{2i+1,0}^m$	$S_{2i+1,0}^m$	$c_{2i,0}^m$	$S_{2i,0}^m$
$c_{2i+1,1}^m$	$S_{2i+1,1}^m$	$c_{2i,1}^m$	$S_{2i,1}^m$

↓

$c_{i,0}^{m+1}$	$S_{i,0}^{m+1}$
$c_{i,1}^{m+1}$	$S_{i,1}^{m+1}$

Before providing the formulae for calculating the sums and carry bits of step m+1, we will demonstrate operation of the conditional sum adder on an example, demonstrating that the adder is suitable equally for addition and subtraction.

<u>Example:</u> n = 12; negative numbers are represented in the 2-complement.

The result of the calculation (see table 2.3) can be interpreted as sum or difference as a function of c_{-1}:

Table 2.3 (Specimen conditional sum addition)

	a	0	0	1	1	0	1	0	0	0	1	1	0	
	b	1	1	0	1	0	1	0	1	1	0	0	1	c_{-1}
step 0		0,1	0,1	0,1	1,0	0,0	1,0	0,0	0,1	0,1	0,1	0,1	0,1	0
		1,0	1,0	1,0	1,1	0,1	1,1	0,1	1,0	1,0	1,0	1,0	1,0	1
1		1 1	1,	0 0	0,	1 0	0,	0 1	0,	1 1	0,	1 1		0
		0 0	1,	0 1	0,	1 1	0,	1 0	1,	0 0	1,	0 0		1
2		0 0	0 0	0,		1 0	0 1	0,		1 1	1 1			0
		0 0	0 1	0,		1 0	1 0	1,		0 0	0 0			1
3		0 0	0 0	0,		1 0	0 1	1 1	1 1					0
		0 0	0 1	0,		1 0	1 0	0 0	0 0					1
4						0 0	0 0	1 0	0 1	1 1	1 1			0
						0 0	0 0	1 0	1 0	0 0	0 0			1

Interpretation options:

1. $c_{-1} = 0$ (addition)

$$w_2(001101000110) = 838$$
$$w_2(110101011001) = -679$$ +
$$w_2(000010011111) = 159$$

2. $c_{-1} = 1$ (subtraction)

$$w_2(001101000110) = 838$$
$$w_2(001010100110) = 678$$ − (*)
$$w_2(000010100000) = 160$$

(*) The bits of addend b have been inverted since we are subtracting.

The carries on the far left of a step need not be calculated since we are applying the 2-complement assuming doubling of the sign of the addends or results. Step 0 is an exception, where $S_{n-1,0}^{0}$ and $S_{n-1,1}^{0}$ are calculated from $c_{n-1,0}^{0}$ and $c_{n-1,1}^{0}$ (see the following formulae).

<u>Formulae for calculating sum or carry bits</u>

Applying the rules of Boolean algebra, we obtain:

A. Step 0 logic

$$c_{i,0}^{0} = a_i \cdot b_i \; ; \quad c_{i,1}^{0} = a_i \cdot b_i \; v \; (a_i \oplus b_i) = a_i \; v \; b_i \; ;$$

$$S_{i,0}^{0} = \overline{S_{i,1}^{0}} \; ; \quad S_{i,1}^{0} = a_i \oplus b_i \oplus 1 = a_i . b_i \; v \; \overline{a_i} . \overline{b_i} = c_{i,0}^{0} \; v \; \overline{c_{i,1}^{0}} \; .$$

B. Calculation of step $m+1$ from step m $(0 \le m \le \lceil \log_2 n \rceil - 1)$

$$\lceil c_{i,0}^{m+1} \mid S_{i,0}^{m+1} \rceil = \begin{cases} \lceil c_{2i+1,0}^{m} \mid S_{2i+1,0}^{m}; S_{2i,0}^{m} \rceil & \text{when } c_{2i,0}^{m} = 0 \\ \lceil c_{2i+1,1}^{m} \mid S_{2i+1,1}^{m}; S_{2i,0}^{m} \rceil & \text{when } c_{2i,0}^{m} = 1 \end{cases}$$

$$= \lceil \overline{c_{2i,0}^{m}} \cdot c_{2i+1,0}^{m} \; v \; c_{2i,0}^{m} \cdot c_{2i+1,1}^{m} \mid \overline{c_{2i,0}^{m}} \cdot S_{2i+1,0}^{m} \; v \; c_{2i,0}^{m} \cdot S_{2i+1,1}^{m}; S_{2i,0}^{m} \rceil$$

$$\lceil c_{i,1}^{m+1} \mid S_{i,1}^{m+1} \rceil = \begin{cases} \lceil c_{2i+1,0}^{m} \mid S_{2i+1,0}^{m}; S_{2i,1}^{m} \rceil & \text{when } c_{2i,1}^{m} = 0 \\ \lceil c_{2i+1,1}^{m} \mid S_{2i+1,1}^{m}; S_{2i,1}^{m} \rceil & \text{when } c_{2i,1}^{m} = 1 \end{cases}$$

$$= \lceil \overline{c_{2i,1}^{m}} \cdot c_{2i+1,0}^{m} \; v \; c_{2i,1}^{m} \cdot c_{2i+1,1}^{m} \mid \overline{c_{2i,1}^{m}} \cdot S_{2i+1,0}^{m} \; v \; c_{2i,1}^{m} \cdot S_{2i+1,1}^{m}; S_{2i,1}^{m} \rceil$$

C. <u>Selection of the result as a function of c_{-1}</u>

Let $S_{0,0}^{k} = t_{n-1}^{0} \ldots t_{0}^{0}$ and $S_{0,1}^{k} = t_{n-1}^{1} \ldots t_{0}^{1}$ be the two alternative results ($k = \lceil \log_2 n \rceil$).

The bits of the sum $S = s_{n-1}\ldots s_0$ are then calculated as follows:

$$s_i = t_i^0 \cdot \overline{c_{-1}} \vee t_i^1 \cdot c_{-1} \qquad (i = 0,\ldots,n-1) .$$

Hence, the first of the two alternatives is used for addition, the second for subtraction. The cycle time can be reduced slightly by inserting this choice in the logic of the previous step. We will not, however, deal further with this.

Analysis of cycle time and costs:

Cycle time. Besides selection of the result, each step of the adder requires passage of one AND or OR gate respectively.

$$\tau_{\text{Cond sum}} = 2 \cdot (\lceil \log_2 n \rceil + 2) .$$

Costs. We will calculate the costs for $n = 2^k$ ($k \in \mathbb{N}$). If n is not a square, the costs can be obtained from the outlay for the next higher square less the outlay for the additional bits.

Step 0 or selection of result: costs 6n each.

Step m ($1 \leq m \leq \log_2 n$):

$\boxed{c_{i,0}^m \mid s_{i,0}^m}$ and $\boxed{c_{i,1}^m \mid s_{i,1}^m}$ are to be calculated, ($i=0,\ldots,\frac{n}{2^m} - 1$).

Half the 2^m sum bits of each group as well as the carry bits (apart from the carry of the group on the far left, which does not need to be calculated) are calculated from the accompanying group carries of step m-1 by two AND gates and one OR gate respectively (costs for each = 6). The remaining half of the sum bits is taken unchanged from step m-1 (see the corresponding formulae). The total costs of the step therefore amount to:

$$\kappa_m = 2 \cdot \underbrace{\frac{n}{2^m} \cdot 2^{m-1} \cdot 6}_{} + \underbrace{2 \cdot (\frac{n}{2^m} - 1) \cdot 6}_{} = 6n + \frac{6n}{2^{m-1}} - 12 .$$

[Costs for: sum bits   carry bits]

$$\kappa_{\text{Cond sum}} = 6n + \sum_{m=1}^{\log_2 n} \kappa_m + 6n = 6n \cdot \log_2 n + 24n - 12 \cdot \log_2 n - 12 .$$

Apart from high speed and relatively low costs, a special advantage of the conditional sum adder is the fact that it requires only AND and OR gates with two inputs each in addition to negation terms. Furthermore, overlapping of several additions (pipeline principle) is possible, since all bits of the adder step are calculated in complete synchrony as a function only of the bits of the immediately preceding step.

2.6 Carry select addition

The conditional sum addition principle can be generalized by combining g (g ≥ 2) groups of one step with a new group of the next higher step. The group size in step i is g^i as against 2^i in the conditional sum addition, whereby the total number of steps is reduced to $1 + \lceil \log_g n \rceil$. A special adder of this kind with a word length n = 100, with group size g = 5 and designed for a maximum target speed with relatively low outlay was specified by Bedrij [Be1]. The adder logic set out in the following generalizes and systematizes these investigations to arbitrary group sizes and word lengths. For simplicity we will dispense with the numerous tricks by which costs can be reduced while maintaining the necessary target speed.

Structure of the carry select adder (group size g):

The formulae for calculating step 0 (group size $g^0 = 1$) correspond to those of the conditional sum addition. The same applies to selection of the sum bits from both alternatives available for the result of the last step.

We shall now describe the logic for calculating the groups $S_{0,0}^{m+1}$ and $S_{0,1}^{m+1}$ (due to the similar structure of all the groups it is sufficient to examine this particular group).

$c_{g-1,0}^{m}$	$S_{g-1,0}^{m}$	$c_{g-2,0}^{m}$	$S_{g-2,0}^{m}$	...	$c_{1,0}^{m}$	$S_{1,0}^{m}$	$c_{0,0}^{m}$	$S_{0,0}^{m}$
$c_{g-1,1}^{m}$	$S_{g-1,1}^{m}$	$c_{g-2,1}^{m}$	$S_{g-2,1}^{m}$	...	$c_{1,1}^{m}$	$S_{1,1}^{m}$	$c_{0,1}^{m}$	$S_{0,1}^{m}$

⇓

$c_{0,0}^{m+1}$	$S_{0,0}^{m+1} \stackrel{def}{=} (S_{g-1,0}'; S_{g-2,0}'; \ldots ; S_{1,0}'; S_{0,0}')$
$c_{0,1}^{m+1}$	$S_{0,1}^{m+1} \stackrel{def}{=} (S_{g-1,1}'; S_{g-2,1}'; \ldots ; S_{1,1}'; S_{0,1}')$

For the g components $S_{i,\alpha}'$ ($\alpha \in \{0,1\}$) of $S_{0,\alpha}^{m+1} = (S_{g-1,\alpha}', \ldots, S_{0,\alpha}')$ we have:

$$S_{i,\alpha}' \in \{S_{i,0}^{m}, S_{i,1}^{m}\} \qquad (i = 0,\ldots,g-1).$$

Which of the two alternatives is to be selected depends on the carries $c_{i-1,\alpha}^{m}, \ldots, c_{0,\alpha}^{m}$. By generalizing the corresponding formulae for the conditional sum addition, we have in particular:

$$S_{0,\alpha}' = S_{0,\alpha}^{m} ;$$

$$S_{1,\alpha}' = S_{1,0}^{m} \cdot \overline{c_{0,\alpha}^{m}} \vee S_{1,1}^{m} \cdot c_{0,\alpha}^{m} ;$$

$$S_{i+1,\alpha}' = S_{i+1,0}^{m} \cdot (\overline{c_{i,1}^{m}} \vee \overline{c_{i,0}^{m}} \cdot \overline{c_{i-1,1}^{m}} \vee \ldots \vee \overline{c_{i,0}^{m}} \cdot \ldots \cdot \overline{c_{1,0}^{m}} \cdot \overline{c_{0,\alpha}^{m}})$$

$$\vee S_{i+1,1}^{m} \cdot (c_{i,0}^{m} \vee c_{i,1}^{m} \cdot c_{i-1,0}^{m} \vee \ldots \vee c_{i,1}^{m} \cdot \ldots \cdot c_{1,1}^{m} \cdot c_{0,\alpha}^{m}) ;$$

$$(i = 1,\ldots,g-2; \alpha \in \{0,1\})$$

$$c_{0,\alpha}^{m+1} = c_{g-1,0}^m \cdot (\overline{c_{g-2,1}^m} \vee \overline{c_{g-2,0}^m} \cdot \overline{c_{g-3,1}^m} \vee \ldots \vee \overline{c_{g-2,0}^m} \cdot \ldots \cdot \overline{c_{1,0}^m} \cdot \overline{c_{0,\alpha}^m})$$

$$\vee\, c_{g-1,1}^m \cdot (\overline{c_{g-2,0}^m} \vee \overline{c_{g-2,1}^m} \cdot \overline{c_{g-3,1}^m} \vee \ldots \vee \overline{c_{g-2,1}^m} \cdot \ldots \cdot \overline{c_{1,1}^m} \cdot \overline{c_{0,\alpha}^m}).$$

These relations (for checking, observe that: $c_{i,0}^m = 1 \Rightarrow c_{i,1}^m = 1$ as well as $c_{i,1}^m = 0 \Rightarrow c_{i,0}^m = 0$) demonstrate that carry select addition combines in itself elements of the conditional sum adder and the carry look ahead adder. The costs for the logic of a group of size g increase with g^3 (see 2.3.2).

Analysis of cycle time and costs:

Cycle time: Two AND and OR gates will be passed respectively in each step when applying the given formulae (where $g > 2$):

$$\tau_{\text{Carry select}} = \begin{cases} 4 + 2 \cdot \lceil \log_2 n \rceil, & g = 2 ; \\ 4 + 4 \cdot \lceil \log_g n \rceil, & g > 2 . \end{cases}$$

Example: $n = 100$; $g = 5$

$$\tau_{\text{Carry select}} = 4 + 4 \times 3 = 16 < \tau_{\text{Cond-sum}} = 4 + 2 \cdot 7 = 18 .$$

Costs: We will determine the outlay for the particular case $n = g^k$ ($k \in \mathbb{N}$). In the same way as for the conditional sum adder, this immediately yields:

$$\kappa_{\text{Carry-select}} = 12n + \sum_{m=1}^{\log_g n} \kappa_m ,$$

where κ_m denotes the outlay for calculating the sum and carry bits, occurring in step m.

Following simple calculation, we have for κ_m:

$$K_m = \frac{n}{\frac{m}{g}} \cdot [\frac{1}{3}g^3 + 3g^2 - \frac{16}{3}g - 4] + 2 \cdot [\frac{n}{\frac{m}{g}}(g^m - g^{m-1}) \cdot 6 + (\frac{n}{\frac{m}{g}} - 1) \cdot 6]$$

$$= \frac{n}{\frac{m}{g}} \cdot [\frac{1}{3}g^3 + 3g^2 - \frac{16}{3}g + 8] + 12n - 12\frac{n}{g} - 12 \ .$$

The costs for the carry select adder are calculated as:

$$K = 12n + 12(n-\frac{n}{g}-1) \cdot \log_g n + (\frac{1}{3}g^3 + 3g^2 - \frac{16}{3}g + 8) \cdot n \cdot \sum_{m=1}^{\log_g n} \frac{1}{g^m}$$

$$= 12n + 12(n-\frac{n}{g}-1) \cdot \log_g n + (\frac{1}{3}g^3 + 3g^2 - \frac{16}{3}g + 8) \cdot (n-1)/(g-1) \ .$$

This can be converted to:

$$K_{\text{Carry select}}$$

$$= 12n + 12(n-\frac{n}{g}-1) \cdot \log_g n + (\frac{1}{3}g^2 + \frac{10}{3}g - 2 + \frac{6}{g-1})(n-1)$$

$$\approx 12n \cdot \log_g n + \frac{n}{3}(g^2 + 10g - 6) \ .$$

2.7 Conclusion, comparison

Table 2.4 combines cycle time and costs for addition of two n digit numbers for the adders discussed in chapter 2. Besides the exlcusive OR adder, adders which calculate the result in several functions (v. Neumann and CSA addition) have not been included since it does not appear reasonable to compare them with others as regards cycle time and costs.

The values in the table demonstrate that, as regards a quick and comparatively cheap addition (with the cost function used by us), the following adders are particularly advantageous:

- 2nd order carry look ahead,
- carry skip with variable group size,
- conditional sum,
- carry select .

Table 2.4

	cycle time	costs ≈	time · costs ≈
c ripple (synchronous)	$2n+2$	$14n$	$28n^2+28n$
c ripple (synchronous)	$2n$	$14n+11$	$28n^2+22n$
c ripple (asynchronous)	$\leq 2 \cdot L+4$	$25n-11$	$50n \cdot L$
complete CLA	5	$\frac{1}{6} \cdot n^3 + 3n^2 + 11\frac{5}{6}n$	$\frac{5}{6}n^3 + 5n^2$
complete CLA	4	$\frac{1}{3} \cdot n^3 + 3n^2 + 8\frac{2}{3}n$	$\frac{4}{3}n^3 + 12n^2$
1st order CLA	$2 \cdot n/g + 3$	$\frac{1}{6} \cdot n \cdot g^2 + \frac{3}{2} \cdot ng$	$\frac{1}{3} \cdot n^2 g + \frac{1}{2} \cdot ng^2$
2nd order CLA ($g=g'$)	$2 \cdot n/g^2 + 7$	$\frac{1}{6} \cdot n \cdot g^2 + \frac{5}{3} \cdot ng$	$\frac{1}{3} \cdot n^2 + \frac{7}{6} \cdot ng^2$
carry skip	$4g + 2n/g - 3$	$15n + 4 \cdot \frac{n}{g}$	$60ng$
c skip (g_{opt})	$4\sqrt{2}\sqrt{n} - 3$	$15n + 4 \cdot \sqrt{2}\sqrt{n}$	$60 \cdot \sqrt{2} \cdot n\sqrt{n}$
c skip (variable)	$4\sqrt{n}$	$15n + 8\sqrt{n}$	$60n \cdot \sqrt{n} + 32n$
conditional sum	$2 \cdot \lceil \log_2 n \rceil + 4$	$6n \cdot \log_2 n + 24n$	$12n \cdot (\log_2 n)^2$
carry select	$4 \cdot \lceil \log_g n \rceil + 4$	$12n \cdot \log_g n + \frac{ng^2}{3}$	$48n \cdot (\log_g n)^2$

n = word length (in bits)
g = group size
L = maximum propagation chain length

3. Multiplication

3.1 Register configuration, notation, overflow problems

In nearly all the procedures presently applied in electronic computers, multiplication is based on a sequence of additions; serial and parallel methods differ with regard to the order in which addition is performed.

In the following we assume that both factors of the product are represented in registers of the same size with length n. Treatment of operands of different length does not give rise to substantial difficulties, but however, a number of different cases should be observed in the microprograms. The following registers are used in particular:

$MD = [MD_{n-1},\ldots,MD_0]$: multiplicand register;
$MQ = [MQ_{n-1},\ldots,MQ_0]$: multiplier register;
$MP = [MP_{n-1},\ldots,MP_0]$: partial product register;
$MH = [MH_{n-1},\ldots,MH_0]$: standby register.

The result of the multiplication is twice as long as the factors; it is usually stored in the coupled register (MP,MQ). The original content of the multiplier register is lost in this organization.

One of the three binary digit codings introduced in chapter 1 will be used as notation. Advantages and disadvantages of the individual codings for multiplication (and/or for division) are as follows:

1. Amount and sign

This is one of the most suitable representations for algorithms of multiplication and division. It is sufficient to multiply or divide the amounts of the operands; the sign of the result is found by comparing signs of the two operands.

2. (d-1)-complement

A negative number represented in the (d-1)-complement can be
converted to a A+S number by inverting all bits except the sign
(and vice versa). This gives rise to the following simple
algorithm (see also 1.3.5):

$$MD_{n-1}MD_{n-2}\ldots MD_0 \qquad MQ_{n-1}MQ_{n-2}\ldots MQ_0 \qquad \text{factors: (d-1)-complement}$$
$$MD_{n-1}MD^*_{n-2}\ldots MD^*_0 \qquad MQ_{n-1}MQ^*_{n-2}\ldots MQ^*_0 \qquad \text{factors: (A+S)}$$

$$MP_{2n-1}MP^*_{2n-2}\ldots\ldots\ldots\ldots\ldots MP^*_0 \qquad \text{product: (A+S)}$$
$$MP_{2n-1}MP_{2n-2}\ldots\ldots\ldots\ldots\ldots MP_0 \qquad \text{product: (d-1)-complement}$$

Application of the (d-1)-complement in the multiplication algorithm
itself is not very suitable (treatment of numerous special cases,
end around carry with addition, see 3.3.4).

3. d-complement

Conversion to a A+S representation or reconversion is, in this case,
more time consuming than with the (d-1)-complement. Nevertheless,
conversion of one or both factors is appropriate in many cases, since
treatment of factors represented in the d-complement is much more
complicated than in A+S representations.

Overflow problems

The product of two n-digit numbers can be stored in a register of length
n with any of the digit codings considered. Doubling of the sign
(d-complement) or insertion of a zero after the sign (A+S) as with
adders is not absolutely necessary in this case; however, the registers
MP and MD must be extended then by one digit, and as regards d-complement notations, a time consuming special case must be considered (see
remarks following microprogram M1). When doubling the sign, these
problems do not occur.

Scaling of the result register (MP,MQ) which is 2n bits long depends on the position of the scaling of the operands:

I. __Integer representation__ (d-complement)

$$-d^{n-1} \le w^{(n,0)}(MD) \, , \, w^{(n,0)}(MQ) \le d^{n-1} -1 \, ;$$

$$-d^{2n-2} + d^{n-1} \le w^{(n,0)}(MD) \cdot w^{(n,0)}(MQ)$$

$$= w^{(2n,0)}(MP,MQ) \le d^{2n-2} \, .$$

The result of the multiplication can be rounded so as to correspond with the representation of the operands only if:

$$MP_{n-1} = MP_{n-2} = \ldots = MP_0 = MQ_{n-1} \, .$$

In this case, the register MQ contains the entire product.

II. __Fixed-point representation $w^{(1,n-1)}$__ (d-complement)

$$-1 \le w^{(1,n-1)}(MD) \, , \, w^{(1,n-1)}(MQ) \le 1 - \frac{1}{d^{n-1}} \, ;$$

$$-1 + \frac{1}{d^{n-1}} \le w^{(1,n-1)}(MD) \cdot w^{(1,n-1)}(MQ)$$

$$= w^{(2,2n-2)}(MP,MQ) \le 1 \, .$$

Agreement between product representation and representation of the operands can be achieved by a shift to the left of (MP,MQ) (the position of the point is shifted also) and by then rounding the register to n digits.

$$
\begin{array}{ccc}
(MP,MQ) & \xrightarrow{\text{SHL + rounding}} & MP^* \\
\downarrow & & \downarrow \\
w^{(2,2n-2)}(MP,MQ) & \xrightarrow{\text{rounding in Q}} & w^{(1,n-1)}(MP^*)
\end{array}
$$

An exception is the product $(-1) \times (-1) = +1$. In this case (and no other) we have $MP_{n-1} \neq MP_{n-2}$.

With regard to floating-point representation, this "overflow" can be eliminated by a shift to the left of the point in (MP, MQ) and simultaneously increasing the exponent by one. This possibility does not exist for fixed-point numbers. This exceptional case cannot occur with regard to representation in the (d-1)-complement or amount and sign.

The algorithm of the multiplication itself is not influenced by the position of the points in the case of operands. Hence we shall restrict ourselves to multiplication of integral numbers for all methods; in all other cases the value of the product is obtained by scaling.

3.2 Serial multiplication with multiplier coding

3.2.1 The microprogram M1

The simplest multiplication procedure consists of n conditional additions and n shifts to the right of the partial product. It can be applied only for non-negative multipliers; the sign of the multiplicand, however, is arbitrary.

Definition 3.1. $P^{(0)} := 0$; $\qquad Q^{(j)} := \dfrac{P^j}{d^j}$ $(j=0,..,n)$;

$P^{(j+1)} := P^{(j)} + d^j \cdot MQ_j \cdot w(MD)$ $\quad Q^{(j)}$ *is denoted as*
$(j=0,..,n-1)$; $\qquad\qquad$ *reduced partial product.*

Result. *For non-negative multipliers* $(MQ_{n-1}=0)$ *we have:*

a. $P^{(n)} = \displaystyle\sum_{i=0}^{n-1} d^i \cdot MQ_i \cdot w(MD) = w(MQ) \cdot w(MD)$;

b. $Q^{(j+1)} = \frac{1}{d} \cdot (Q^{(j)} + MQ_j \cdot w(MD))$.

Calculation of $Q^{(j+1)}$ from $Q^{(j)}$ is carried out by a conditional addition and subsequent shift to the right.

The process for the algorithm (calculation of $Q^{(1)},\ldots,Q^{(n)}$) is described by the following microprogram (base d = 2):

Microprogram M1

0 : MD := Mnd; MQ := Mier; MP := 0; U := MQ_{n-1}; Z := n;
0*: if U = 1 then MQ := $\overline{MQ}$ + 1;
1 : Z := Z-1; if MQ_0 = 1 then MP := MP + MD ;
2 : SHR(MP,MQ); if Z > 0 then goto 1;
2*: if U = 1 then (MP,MQ) := $\overline{(MP,MQ)}$ + 1;
3 : END .

Remarks. a. SHR(A) denotes a shift to the right of the register A; here, the sign is shifted subsequently; the extreme right position of A is lost, i.e.: $[a_{n-1}a_{n-2}\cdots a_0] \xrightarrow{SHR(A)} [a_{n-1}a_{n-1}a_{n-2}\cdots a_1]$

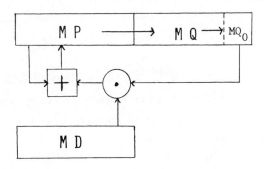

Fig. 3.1

b. The non-integer part of $Q^{(j)}$ is shifted to the no longer necessary part of the register MQ; hence, additions apply only to a register of length n. Since MQ is shifted also, MQ_0 always contains the "real" digit of the multiplier (see fig. 3.1).

c. Overflow situations regarding partial product ($MP_{n-1} \neq MP_{n-2}$ after the addition in function 1) are recognized by doubling of the sign and cancelled by the shift to the right in function 2.

d. Functions 0* and 2* are cancelled with A+S-notation. Conversion to a positive multiplier (MQ := $\overline{MQ}$ + 1) is impossible when MQ = [1 0...0] ; this case cannot occur with doubling of the sign; hence we avoid considering an inconvenient special case.

e. After cycling of the microprogram, we have:

$$w((MP,MQ)) = \begin{cases} Q^{(n)} & \text{when point at end of MQ;} \\ d^n \cdot Q^{(n)} = P^{(n)} & \text{when point at end of MP.} \end{cases}$$

f. The two factors can be of different length; we shall not extend the microprogram to this case since the instructions of the new program would have to refer to parts of registers. Furthermore, in most cases, both factors are stored in registers of the same length.

3.2.2 Direct treatment of negative multipliers in the d-complement

We can expand microprogram M1 so that negative multipliers represented in the d-complement do not require to be conformed first into positive ones. It is seen that:

$$|w(MQ)| = \begin{cases} \sum_{i=0}^{n-2} MQ_i \cdot d^i & \text{when } MQ_{n-1} = 0; \\ \sum_{i=0}^{n-2} \overline{MQ_i} \cdot d^i + 1, & \text{otherwise .} \end{cases}$$

In the special case d = 2, we have:

$$|w(MQ)| = \sum_{i=0}^{n-2} [MQ_i \oplus MQ_{n-1}] \, 2^i + MQ_{n-1} \, .$$

We thus obtain for the product:

$$w(MD) \cdot w(MQ) = (1 - 2 \cdot MQ_{n-1}) \cdot [\sum_{i=0}^{n-2} (MQ_i \oplus MQ_{n-1}) 2^i + MQ_{n-1}]$$
$$\cdot \, w(MD) \, .$$

The new microprogram requires only a slight change as against M1 (change in MP and MQ digits).

Microprogram M2:
===

0 : MD := Mnd; MQ := Mier; U := MQ_{n-1}; Z := n;
0*: MP_j := $MD_j \cdot MQ_{n-1}$; MQ_j := $MQ_j \oplus MQ_{n-1}$
    (j = 0, ... , n-1);

    [continuing as for microprogram M1]

Examples:
===

$$n = 8, \quad d = 2 \, .$$

1) MD = 00111111;
    MQ = 00100111;

    w(MD) = 63;
    w(MQ) = 39.

2) MD = 00111110;
    MQ = 11001111;

    w(MD) = 62;
    w_2(MQ) = − 49.

1)

MP	MQ	
00000000	0010011<u>1</u>	
+ 00111111		
00111111		ADD
00011111	1001001<u>1</u>	SHR
+ 00111111		
01011110		ADD
00101111	0100100<u>1</u>	SHR
+ 00111111		
01101110		ADD
00110111	0010010<u>0</u>	SHR
00011011	1001001<u>0</u>	SHR
00001101	1100100<u>1</u>	SHR
+ 00111111		
01001100		ADD
00100110	0110010<u>0</u>	SHR
00010011	0011001<u>0</u>	SHR
00001001	1001100<u>1</u>	SHR

w(MP,MQ) = 2457 = 63 x 39.

2)

MP	MQ	
00111110	0011000<u>0</u>	
00011111	0001100<u>0</u>	SHR
00001111	1000110<u>0</u>	SHR
00000111	1100011<u>0</u>	SHR
00000011	1110001<u>1</u>	SHR
+ 00111110		
01000001		ADD
00100000	1111000<u>1</u>	SHR
+ 00111110		
01011110		ADD
00101111	0111100<u>0</u>	SHR
00010111	1011110<u>0</u>	SHR
00001011	1101111<u>0</u>	SHR
11110100	00100010	←

correction, since U=1 ⤴

w_2(MP,MQ) = -3038 = 62 x (-49).

3.2.3 Acceleration of multiplication by carry save addition

The microprograms M1 and M2 are very time consuming since the duration of a function must accord with the (maximum) cycle time for one addition. A decisive reduction in the function time can be achieved by applying a carry save adder. The addition carries are stored in a standby register MH and processed in the next function. The contents of the MP and MH register are added following the n-th multiplication cycle.

Microprogram M3

0 : $MD := Mnd$; $MQ := Mier$; $MH := 0$; $U := MQ_{n-1}$; $Z := n$;
0*: $MP_j := MD_j \cdot MQ_{n-1}$; $MQ_j := MQ_j \oplus MQ_{n-1}$ ($j=0,...,n-1$);
1 : $Z := Z-1$; $MP_j := MH_j \oplus MP_j \oplus MQ_0 \cdot MD_j$;
  $MH_j := MH_j \cdot MP_j \vee MH_j \cdot MQ_0 \cdot MD_j \vee MP_j \cdot MQ_0 \cdot MD_j$; ($j=0,..,n-1$)
2 : $SHR(MP,MQ)$; if $Z > 0$ then goto 1;
3 : if $MH \neq 0$ then [$MP_j := MH_j \oplus MP_j$; $MH_{j+1} := MP_j \cdot MH_j$ ($j=0,..,n-2$);
  $MH_0 := 0$; goto 3];
3*: if $U \neq 0$ then [$(MP,MQ):=(\overline{MP},\overline{MQ})$; $(MD,MH) := 00...01$] else goto 5;
                              A   B
4*: if $B \neq 0$ then [$A_j := A_j \oplus B_j$ ($j=0,..,2n-1$);
  $B_{j+1} := A_j \cdot B_j$ ($j=0,..,2n-2$); $B_0 := 0$; goto 4*] ;
5 : END .

Von Neumann additions are carried out at two positions of this program (function 3 and function 4*).

3.2.4 Multiplier zero and unity shifts

(Booth method [Le3])

If the multiplier contains a zero block of length k, we can accelerate multiplication by shifting the partial product through k digits. Since

$$w(0...01...10...0) = \sum_{i=u}^{v} 2^i = 2^{v+1} - 2^u$$
$$\underset{v}{|}\underset{u}{|}$$

we can replace v−u+1 additions of the multiplicand (associated with one shift in each case) by subtraction (on position u) and addition (on position v+1) with a unity block in MQ. Hence, arithmetic operations are necessary only at 01- or 10-multiplier changes.

The Booth method (see table 3.1) based on this can be applied
for all binary digit codings described. The individual variants
differ in execution of addition or subtraction and in occupation of
the register position MQ_{-1}.

Table 3.1 (Booth method)

MQ_i	MQ_{i-1}	operation
0	1	ADD; shift
1	0	SUB; shift
0	0	---; shift
1	1	---; shift

Microprogram M4 (Booth method; 2-complement notation)

An MQ_{-1} digit which must be occupied by 0 in the 2-complement, is
added to $MQ = MQ_{n-1}...MQ_0$. Besides w(MD), we also require -w(MD)
since subtractions are to be performed also. Hence, first we store
$-w(MD) = w(\overline{MD})+1$ in a standby register MH; observe that MD = [10...0]
is impossible due to doubling of the sign.

0 : MD := Mnd; MQ := Mier; MQ_{-1} := 0; MP := 0; Z := n;
1 : MH := $\overline{MD}$ + 1;
2 : Z := Z-1; if (MQ_0, MQ_{-1}) = (0,1) then MP := MP + MD;
             if (MQ_0, MQ_{-1}) = (1,0) then MP := MP + MH;
3 : SHR(MP,MQ); if Z > 0 then goto 2;
4 : END .

Remarks. a. The Booth method provides the simplest example of a
more general technique (multiplier coding) as considered in 3.3 .
Proof that this method is correct for positive as well as negative
multipliers is demonstrated by the result obtained for the multiplier coding (see 3.3.2). Note that, contrary to conventional

multiplication algorithm, microprogram M4 does not require correction of the result as a function of the sign of the multiplier or product.

b. Additions in function 2 of M4 can be performed overlapping employing a carry save adder. (See 3.2.3).

c. If shifts are possible over more than one digit, we can shift over entire zero or unity blocks. Hence, the number of multiplication cycles is reduced on average by approximately a factor 2.6 (see [Fr1]; here, the corresponding result for division is derived.) The related microprogram M5 is based on an additional circuit which determines the length of the block to be shifted over.

<u>Microprogram M5 (variable shifts of random size)</u>

0 : MD := Mnd; MQ := Mier; MQ_{-1} := 0; MP := 0; Z := n;
1 : MH := $\overline{MD}$ + 1;
2 : <u>if</u> (MQ_0, MQ_{-1}) = (0,1) <u>then</u> MP := MP + MD;
　　<u>if</u> (MQ_0, MQ_{-1}) = (1,0) <u>then</u> MP := MP + MH;
　　j := min(L,Z); [L is the length of the present 0- or 1-block:
　　　　　　　　L = k ⇔ $(MQ_{k-1} = \ldots = MQ_0; MQ_k \neq MQ_{k-1})$];
3 : SHR(MP,MQ) over j digits; Z := Z-j;
4 : <u>if</u> Z > 0 <u>then</u> <u>goto</u> 2;
5 : END .

The Booth method is ineffective for blocks of length 1. The number of arithmetic operations to be performed can be reduced by means of an additional logic which recognizes blocks of length 1 and performs simpler calculation. The simplified method operates as follows ($\tilde{1}$:= -1):

MQ =   ...00100...   ...0111111101111110...

Booth method   ...011̃00...   ...1000001̃10000010̃...

simplified method   ...00100...   ...10000001̃0000010̃...

Examples of the Booth method

1. MD = 11010010011 ;   2. MD = 00001101011 ;
    MH = 00101101101 ;   MH = 11110010101 ;
    MQ = 11110011000 ;   MQ = 11110001111 ;

    $w_2(MD) = -365$; $w_2(MD) = + 107$;
    $w_2(MQ) = -104$. $w_2(MQ) = - 113$.

MP	MQ	MQ_{-1}		MP	MQ	MQ_{-1}	
00000000000	11110011000	0		00000000000	11110001111	0	
00000000000	00011110011	0	SHR(3)	+ 11110010101			Add MH
+ 00101101101			Add MH	11110010101	11110001111	0	
00101101101	00011110011	0		11111111001	01011111000	1	SHR(4)
00001011011	01000111100	1	SHR(2)	+ 00001101011			Add MD
+ 11010010011			Add MD	00001100100	01011111000	1	
11011101110	01000111100	1		00000001100	10001011111	0	SHR(3)
11110111011	10010001111	0	SHR(2)	+ 11110010101			Add MH
+ 00101101101			Add MH	11110100001	10001011111	0	
00100101000	10010001111	0		11111111010	00011000101	1	SHR(4)
00000010010	10001001000	1	SHR(4)				(*)

$w_2(MP,MQ) = 37960$ (*): $j=\min(L,Z) = \min(5,4)=4$;
             $= (-365) \times (-104)$. $w_2(MP,MQ) = -12091$
                                                     $= 107 \times (-113)$.

The respective shift size j is shown in brackets; j must not be larger
than the existing Z count (see example 2) .

3.3 Multiplier coding

Multiplier coding is aimed at reducing the number of additions or subtractions in multiplication by using suitable multiples of the multiplicand. The earliest example of this technique was the Booth method.

If V times the multiplicand ($0 \leq V \leq 2^h-1$) is available, we can combine h cycles of the multiplication algorithm into a single one (multiplication with base 2^h). Multiplier coding achieves the same by using smaller multiples V of the multiplicand ($-2^{h-1} \leq V \leq +2^{h-1}$).

3.3.1 Multiplier coding for non-negative multipliers (groups each of h+1 bits)

Let $MQ = [MQ_{n-1},\ldots,MQ_0]$, $MQ_{n-1} = 0$, i.e. $w(MQ) \geq 0$.

Starting from the right, we divide the multiplier into groups of $h(h \in \mathbb{N})$ bits respectively. Without restriction we assume that h is a divisor of n (otherwise special treatment for the group on the far left). Group i together with the next bit of group (i-1) on the right, provides a multiple k_{hi} of the multiplicand; the h multiplication cycles necessary for this group are replaced by addition of $k_{hi} \cdot 2^{hi} \cdot w(MD)$.

Definition 3.2. *The multiple k_{hi} is defined as follows:*

$$k_{hi} = k_{hi}(MQ_{(i+1)h-1},\ldots,MQ_{ih}, MQ_{ih-1})$$
$$:= -2^{h-1} MQ_{(i+1)h-1} + \sum_{j=0}^{h-2} MQ_{ih+j} \cdot 2^j + MQ_{ih-1} \cdot$$

We will now demonstrate that with this choice of k_{hi} the relation

$$w(MD) \cdot w(MQ) = \sum_{i=0}^{n/h-1} [k_{hi} \cdot w(MD)] \cdot 2^{hi}$$

is satisfied. It is sufficient to prove:

$$w(MQ) = \sum_{i=0}^{n/h-1} k_{hi} \cdot 2^{hi} .$$

This is verified for non-negative MQ in the following lemma:

Lemma 3.1. *Let $w(MQ) \geq 0$ and $MQ_{-1} := MQ_{n-1} = 0$; now:*

a. $-2^{h-1} \le k_{hi} \le 2^{h-1}$; b. $w(MQ) = \sum_{i=0}^{n/h-1} k_{hi} \cdot 2^{hi}$;

c. $k_{hi}(\overline{MQ}_{(i+1)h-1}, \ldots, \overline{MQ}_{ih-1}) = -k_{hi}(MQ_{(i+1)h-1}, \ldots, MQ_{ih-1})$.

Proof. Part a. is trivial. Since $MQ_{n-1} = 0$, we have for proposition b:

$$w(MQ) = \sum_{i=0}^{n-1} MQ_i \cdot 2^i = \sum_{i=0}^{n/h-1} [(\sum_{j=0}^{h-1} MQ_{ih+j} \cdot 2^j) \cdot 2^{hi}]$$

$$= \sum_{i=0}^{n/h-1} [(-MQ_{(i+1)h-1} \cdot 2^{h-1} + \sum_{j=0}^{h-2} MQ_{ih+j} \cdot 2^j) \cdot 2^{hi}]$$

$$+ \underbrace{\sum_{i=0}^{n/h-1} 2 \cdot MQ_{(i+1)h-1} \cdot 2^{h-1} \cdot 2^{hi}}_{(*)}.$$

Since $MQ_{n-1} = MQ_{-1} = 0$, (*) can be converted to:

$$(*) = \sum_{i=0}^{n/h-1} MQ_{(i+1)h-1} \cdot 2^{h(i+1)} = \sum_{i=1}^{n/h} MQ_{ih-1} \cdot 2^{hi} = \sum_{i=0}^{n/h-1} MQ_{ih-1} \cdot 2^{hi}.$$

Hence:

$$w(MQ) = \sum_{i=0}^{n/h-1} [(-MQ_{(i+1)h-1} \cdot 2^{h-1} + \sum_{j=0}^{h-2} MQ_{ih+j} \cdot 2^j + MQ_{ih-1}) \cdot 2^{hi}].$$

Proposition b is thus demonstrated. Proposition c. is verified similarly.

The formulae of the multiplier coding yield the following microprogram:

Microprogram M6 (multiplier coding in groups of h+1 bits)

0 : MD := Mnd; MQ := Mier; MP := 0; Z := n; MQ_{-1} := 0;
1 : k := $-2^{h-1} \cdot MQ_{h-1}$

$+ \sum_{j=0}^{h-2} MQ_j \cdot 2^j + MQ_{-1}$;

2 : MH := k · MD;
3 : Z := Z-h; MP := MP + MH;
4 : SHR(MP,MQ) over h digits; if Z > 0 then goto 1;
5 : END .

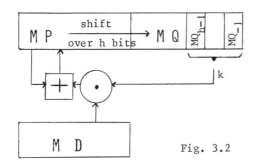

Fig. 3.2

In addition to the sign of the multiplier, application of this microprogram depends also on representation of the negative multiple of the multiplicand (see following discussion).

A. 2-complement notation

All intermediate results are represented in the 2-complement. It is demonstrated in 3.3.2 that M6 also provides the correct result for negative multipliers.

B. Notation by amount and sign

M6 is applicable in all cases. Partial products and multiples of the (value of the) multiplicand can become negative, and the end result is non-negative in every case. Hence the 2-complement can be used for representation of the intermediate results.

C. 1-complement notation

This method of representation is completely unsuitable. It can readily be seen that all additions (function 3 of microprogram M6) have to be extended to the full length of the partial product. Hence, the length of register MH would have to be variable. Apart from greater expenditure for these additions, there would also be longer cycle time through the end around carry. We will demonstrate these problems on an example in 3.3.4 . As frequently mentioned, conversion to coding by amount and sign is advantageous with representation of the factors in the 1-complement.

We will deal with overflow problems in 3.3.3 . Before providing examples of multiplier coding, we will extend M6 to negative multipliers.

3.3.2 Multiplier coding for any multipliers (1- and 2-complement notation)

Lemma 3.2.

If $MQ_{-1} := \begin{cases} MQ_{n-1} & \text{1-complement representation} \\ 0 & \text{2-complement representation;} \end{cases}$

then the formulae for multiplier coding apply <u>independently</u> of the sign of the multiplier.

Proof. We have to prove $w(MQ) = \sum_{i=0}^{n/h-1} k_{hi} \cdot 2^{hi}$.

We have demonstrated this for $MQ_{n-1} = 0$ in 3.3.1 . If $MQ_{n-1} = 1$ ($w(MQ) \leq 0$), then:

$$w(MQ) = -[-w(MQ)] = -[w(\overline{MQ}) + \alpha] \text{ with } \alpha = \begin{cases} 0 & \text{1-complement} \\ 1 & \text{2-complement.} \end{cases}$$

$$= -[\sum_{i=0}^{n/h-1} k_{hi}(\overline{MQ}_{(i+1)h-1}, \ldots, \overline{MQ}_{ih-1}) \cdot 2^{hi} + \alpha]$$

$$= \begin{cases} -[-\sum_{i=0}^{n/h-1} k_{hi} \ (MQ_{(i+1)h-1},\ldots,MQ_{ih-1}) \cdot 2^{hi}], \\ \qquad \text{when } MQ_{-1} = MQ_{n-1} = 1 \quad (1\text{-complement}) \\ -[-\sum_{i=0}^{n/h-1} k_{hi} \ (MQ_{(i+1)h-1},\ldots,MQ_{ih-1}) \cdot 2^{hi} - 1 + 1] \\ \qquad \text{when } MQ_{-1} = 0 \quad (2\text{-complement}) \ . \end{cases}$$

The last equation is obtained by substituting MQ_{-1}:

$$k_0(\overline{MQ_{h-1}},\ldots,\overline{MQ_0},0) \qquad = -k_0(MQ_{h-1},\ldots,MQ_0,1)$$
$$= -[k_0(MQ_{h-1},\ldots,MQ_0,0) + 1] = -k_0(MQ_{h-1},\ldots,MQ_0,0) - 1 \ .$$

Remark: The Booth method is a special case (h=1) of multiplier coding. In this case we have $k_i := -MQ_i + MQ_{i-1}$.

Hence it is not necessary to change microprogram M6 if the multiplier is negative and if the 2-complement is used as notation.

All multiples k_{hi} (i=0,...,n/h-1) can be calculated in parallel; this is important for parallel methods of multiplication. There are various methods of the multiplier coding which do not permit parallel calculation of all multipliers.

3.3.3 Overflow problems

Since $k_{hi} \in [-2^{h-1} : 2^{h-1}]$ and assumed doubling sign, the registers MP and MD will be extended by h-1 (sign-) bits (2-complement notation). As regards the 1-complement or amount and sign, an extension by h-2 bits would be sufficient. The following example demonstrates an extreme case where h-1 additional digits are necessary in 2-complement notation.

Example: h = 3   (2-complement notation)

	MP	MQ	MQ_{-1}
MQ = 110100100 w_2(MQ) = - 92	00 000000000	110100100	0
MD = 110000000 w_2(MD) = -128	+ 01 000000000	-4	
= 11 110000000	01 000000000		
(extension of MD by	00 001000000	000)110100	1
h-1=2 digits)	+ 00 110000000	-3	
	00 111000000		
-MD = 00 010000000	00 000111000	000000)110	1
-4MD = 01 000000000	+ 00 010000000	-1	
-3MD = -2MD - MD	00 010111000		
= 00 110000000 .	00 000010111	000000000	1

w_2(MP,MQ) = 11776 = (-92) x (-128).

3.3.4 Examples

A. Parallel calculation of the multiples k_{hi}

h=4 : $k_{4i} = -8MQ_{4i+3} + 4MQ_{4i+2} + 2MQ_{4i+1} + MQ_{4i} + MQ_{4i-1} \in [-8:+8]$;

h=3 : $k_{3i} = -4MQ_{3i+2} + 2MQ_{3i+1} + MQ_{3i} + MQ_{3i-1} \in [-4,+4]$;

h=2 : $k_{2i} = -2MQ_{2i+1} + MQ_{2i} + MQ_{2i-1} \in [-2,+2]$.

In each of the following examples we will use the same negative multiplier represented in the 2-complement. With regard to 1-complement coding, the multiple k_0 standing on the far right with negative multipliers has a value which is higher by 1 . All other multiples remain unchanged.

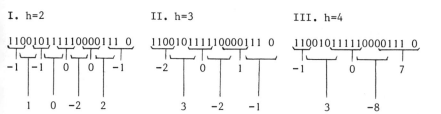

It is evident that the sizes of the multiplier groups can differ. In this case we call the marginal group on the left a size 2 group. We would obtain the same result by extending this group by two additional sign bits.

B. Application of microprogram M6 (serial multiplication)

1. $h=3$, $n=10$ (special treatment of the last group)

```
MD   =  0001101000    | w₂(MD) = 104
     =  000001101000  | (extended by
MQ   =  0011101101    | two bits);
-MD  =  111110011000  | w₂(MQ) = 237
-2MD =  111100110000
-3MD =  111011001000
+4MD =  000110100000
```

Addition is necessary for calculating ± 3MD.

The last group of the multiplier consists of one single bit only. Hence, when shifting for the last time, there is a shift only by one digit.

MP	MQ	MQ₋₁
000000000000	0011101<u>101 0</u>
+111011001000	−3
111011001000	
111111011001	000<u>0011 101 1</u>
+111100110000	−2
111100001001	
111111100001	00100<u>0</u><u>0011 1</u>
+000110100000	+4
000110000001	
000000110000	0010010<u>00</u>0 0
	0
000000011000	0001001000 ∅

The factors and final result have the same value for each of the three digit codings dealt with:

$w(MQ) \times w(MD) = 104 \times 237 = 24648 = w(MP,MQ)$.

The calculation is performed similarly when coding the factors by amount and sign and representing negative intermediate results by the 2-complement.

2. $h=2$, $n=10$, 2-complement notation

				MP	MQ	MQ_{-1}
MD	= 1100010011	$w_2(MD) = -237$				
	= 11100010011	(extended by		00000000000	1110011000	0
MQ	= 1110011000	1 bit);				0
2MD	= 11000100110	$w_2(MQ) = -104$		00000000000	0011100110	0
−MD	= 00011101101			+00111011010		−2
−2MD	= 00111011010			00111011010		
				00001110110	1000111001	1
(all necessary multiples				+11000100110		+2
can be obtained by negating				11010011100		
and/or one shift).				11110100111	0010001110	0
				+00111011010		−2
$w_2(MP,MQ)$	= 24648			00110000001		
	= $w_2(MD) \times w_2(MQ)$.			00001100000	0100100011	1
						0
				00000011000	0001001000	1

3. $h=3$, $n=10$, 1-complement notation

This example demonstrates the problems occurring when multiplying in the 1-complement (variable length of additions, end around carry etc.) .

MD   =   111100101११

	MP	MQ	MQ_{-1}

MD   =   111100101 11
MQ   =   110001001 0
−MD   =   000011010 00
2MD   =   111001011 11
    +   111100101 11
        111011000 110
3MD   =   110110001 11
−4MD =   001101000 00

$w_1(MD) = -104$
$w_1(MQ) = -237$
$w_1(MD) \times w_1(MQ) = 24648$
$\quad = w_1(MP, MQ)$.

```
                    MP           MQ        MQ₋₁
              00000000000   1100010010  1
              +11011000111              +3
              11011000111
              11111011000   1111100010  0
              +11100101111  111         +2
              111100001000  110
              11100001000   111
              11111100001   0001111100  0
              +00110100000  000000      -4
              100110000001  000111
              00110000001   001000
              00000110000   0010010001  1
                                        0
              00000011000   0001001000  ∅
```

With 1-complement notations, the multiples $k_{hi} \times$ MD must be extended to the full length of the partial product (adding i x h signs). MP is extended by h−2 digits only, since the special case possible in the 2-complement (see 3.3.3) cannot occur here. Transition to representation by amount and sign is much simpler; in this case, calculation would be performed as in example 1.

3.3.5 Circuit for calculating $k_{2i} \times$ MD

Multiplier coding is used mainly for group size h=2, since larger values of h require further additions. For application of multiplier coding with parallel multiplications, all multiples of the multiplicand should be calculated at maximum speed. All $k_{2i} \times$ MD can be calculated by a simple circuit in four logical steps; the 2-complement notation is taken as base, since investigations so far have shown that this coding should be applied for representing intermediate results.

Table 3.2

MQ_{2i+1}	MQ_{2i}	MQ_{2i-1}	k_{2i}
0	0	0	0
0	0	1	1
0	1	0	1
0	1	1	2
1	0	0	-2
1	0	1	-1
1	1	0	-1
1	1	1	0

Since

$$MQ_{2i+1} = \begin{cases} 0 \Rightarrow k_{2i} \geq 0 \\ 1 \Rightarrow k_{2i} \leq 0 \end{cases}$$

we obtain from table 3.2 the relationship:

$$k_{2i} \times MD = k_{2i} \cdot [MD_{n-1},\ldots,MD_0]$$
$$= |k_{2i}| \cdot ([\alpha_{n-1},\ldots,\alpha_0] + [0,\ldots,0,MQ_{2i+1}]),$$

where $\alpha_j := MD_j \oplus MQ_{2i+1}$ is used for brevity.

Considering that representation of $k_{2i} \cdot MD$ may be longer by one bit than for MD (doubling of the sign), we have:

$$k_{2i} \times MD$$
$$= \begin{cases} [\alpha_{n-1},\alpha_{n-1},\alpha_{n-2},\ldots,\alpha_0] + [0,\ldots,0,0,MQ_{2i+1}] & \text{if } |k_{2i}| = 1 \\ [\alpha_{n-1},\alpha_{n-2},\alpha_{n-3},\ldots,\alpha_0,0] + [0,\ldots,0,MQ_{2i+1},0] & \text{if } |k_{2i}| = 2 \\ [\ 0,\ 0,\ 0,\ldots,0\ ] & \text{if } |k_{2i}| = 0 \end{cases}$$

$$= [\alpha_{n-1} \cdot (\beta_1 \vee \beta_2), \alpha_{n-1} \cdot \beta_1 \vee \alpha_{n-2} \cdot \beta_2, \ldots, \alpha_1 \cdot \beta_1 \vee \alpha_0 \cdot \beta_2, \alpha_0 \cdot \beta_1]$$
$$+ [\quad 0 \quad , \quad 0 \quad , \ldots, MQ_{2i+1} \cdot \beta_2, MQ_{2i+1} \cdot \beta_1]$$

where: $\beta_r = 1 \stackrel{\text{def}}{\Leftrightarrow} |k_{2i}| = r \quad (r = 1,2);$

i.e. $\beta_1 = MQ_{2i} \oplus MQ_{2i-1} = \overline{MQ}_{2i} \cdot MQ_{2i-1} \vee MQ_{2i} \cdot \overline{MQ}_{2i-1};$

$\beta_2 = MQ_{2i+1} \cdot \overline{MQ}_{2i} \cdot \overline{MQ}_{2i-1} \vee \overline{MQ}_{2i+1} \cdot MQ_{2i} \cdot MQ_{2i-1} \cdot$

Cycle time and costs of the respective circuit are calculated as:

$\tau = 4$ logical steps;

$\kappa = 6n + 4n + 2n + 18 = 12n + 18$.

The total costs for all $\frac{n}{2}$ multiples are:

$\kappa_{total} = 6n^2 + 9n$.

3.3.6 Investigation of effort and cycle time for multiplier coding with group size h

In order to apply multiplier coding, the k_{hi} multiple of the multiplicand $(-2^{h-1} \leq k_{hi} \leq +2^{h-1})$ has to be calculated. For $h \leq 2$, all multiples can be obtained from MD by shifts and/or negation. For $h \geq 3$, however, $2^{h-2}-1$ multiples of the multiplicand (namely the $3,5,7,\ldots,(2^{h-1}-3)$, $(2^{h-1}-1)$ multiple of MD) are necessary which cannot be obtained in this way. This requires additional time; for large values of h, the time saving (reduction in the number of cycles (loops) from n to n/h) is generally lost because of this. For h=3, multiplier coding is often ineffective due to availability of the multiple $\pm 3 \cdot$ MD.

3.4 Unclocked and parallel methods of multiplication

3.4.1 The multiplication matrix M_0

With serial methods of multiplication, the addition cycles occur in sequence; here, each addition requires at least one function. By performing several additions simultaneously and/or by avoiding division into functions, the speed of multiplication can be essentially increased. The costs are, of course, higher than with serial methods.

For designing and analysing fast, unclocked multiplications, all the addends occurring (multiplication is based on a sequence of additions) are set up in one <u>multiplication matrix M_0</u>. Apart from the notation chosen, the form of M_0 also depends on whether the multiplier has been coded or not.

<u>Examples:</u>

a. <u>Both factors non-negative</u>

$MD = [MD_{k-1},\ldots,MD_0]$, $MQ = [MQ_{n-1},\ldots,MQ_0]$ with $MD_{k-1} = MQ_{n-1} = 0$;

$$M_0 = \begin{bmatrix} 0 & a_{k-1,0} & a_{k-2,0} & \cdots\cdots\cdots & a_{0,0} \\ & a_{k-1,1} & a_{k-2,1} & \cdots\cdots\cdots & a_{0,1} \\ a_{k-1,n-1} & \cdots\cdots\cdots & a_{0,n-1} & & 0 \end{bmatrix}$$

where $a_{i,j} := MD_i \cdot MQ_j \in \{0,1\}$.

For this purpose we apply the following, simpler notation:

$$M_0 = \begin{bmatrix} 0 & \cdots \\ & \cdots \\ \cdots & 0 \end{bmatrix} \Big\} n$$

$\longmapsto k \longmapsto$

One dot designates one position of matrix M_0, which can take 1.

All elements of matrix M_0 can be calculated simultaneously in one logical step. For this we require n x k AND gates with two inputs each.

The number of lines and columns can be reduced by disregarding the signs (MD_{k-1} and MQ_{n-1}).

b. <u>Non-negative multiplier, random multiplicand sign</u>

With regard to 2-complement representation, the top left triangle of M_0 must take the sign of the multiplicand.

The bottom right triangle of M_0 must also be occupied in the 1-complement.

$$M_0 = \begin{bmatrix} \cdots\cdots\cdots\cdots\cdots\cdots \\ \quad \quad \quad 0 \end{bmatrix}$$

Since the effort for adding the M_0 lines depends on the filling density of the matrix, notation by <u>amount and sign</u> (matrix M_0 see a.) is <u>the most suitable coding</u> for parallel methods of multiplication.

c. <u>Multiplier coding with group size h > 1</u>

The same number of M_0 lines corresponds to the n/h multiples of the multiplicand. All lines must be filled with the k_{hi} x MD sign to the left (2-complement) or to the right (1-complement). Representation of a coded multiple of the multiplicand is longer than that for MD by up to h-1 bits (see example 3.3.3). In the 2-complement, M_0 contains an additional line which is partially filled if the matrix contains multiples k_{hi} x MD with $k_{hi} < 0$ (addition of a one when forming a negative multiple of a multiplicand); all these further additions can be combined in one line.

For calculation of M_0 by means of a circuit we can use the circuit quoted in 3.3.5 for group size h=2. It can readily be demonstrated that, even with greater values of h, 4 logical steps of cycle time steps are sufficient always for forming the multiples and that the effort necessary for this is greater than 6n x k. As compared with the uncoded version of M_0 (see a.), we have an effort more than three times greater and a cycle time greater by three logical steps; however, do not forget that the total cycle time can diminish due to reduction of the lines from n to $\frac{n}{h} + 1$.

3.4.2 Reduction of M_0 (addition of lines)

Most multiplication methods are intended for numbers of equal length, represented by amount and sign. The original matrix has the following form:

$$M_0 = \begin{bmatrix} & \cdots\cdots \\ 0 & \cdots\cdots \\ & \cdots\cdots \\ & \cdots\cdots \\ \cdots\cdots & 0 \\ \cdots\cdots & \end{bmatrix} \Big\} n$$

$$\longmapsto n \longmapsto$$

(both factors consist of n bits and of the sign, not shown here).

As regards all methods, M_0 is reduced by (3,2)- or (2,2)-counters (fulladders or halfadders) as well as by OR gates via matrices $M_1, M_2, \ldots$ to a matrix M_E, which consists of two lines only. The individual methods differ from each other in the manner of reduction from M_0 to M_E.

The effort is influenced by the choice of the last adder (for addition of the two M_E lines). Since we wish to measure the costs for the number of (3,2)- or (2,2)-counters necessary, we select the carry ripple adder containing only these two types of modules when comparing the costs.

The following demonstrates a simplified representation of matrix M_0 and the intermediate matrices M_i $(1 \leq i \leq E)$:

$$[M_i] := [s_{2n-1}^{(i)}, s_{2n-2}^{(i)}, \ldots, s_1^{(i)}, s_0^{(i)}].$$

$s_j^{(i)}$ denotes the <u>maximum value</u> of the sum of elements in column j. For example, we have:

$$[M_0] = [0,1,2,3,\ldots,n-1,n,n-1,\ldots,2,1];$$

$$[M_E] = [s_{2n-1}^{(E)}, \ldots, s_0^{(E)}] \quad \text{with} \quad 1 \leq s_i^{(E)} \leq 2 \quad (0 \leq i \leq 2n-1);$$

$$[M_{E+1}] = [1,1,\ldots,1] \quad \text{(product "matrix")}.$$

For calculating M_{i+1} from M_i, $f_r^{(i)}$ fulladders and $h_r^{(i)}$ halfadders are entered in columns r $(0 \leq r \leq 2n-1)$ (in exceptional cases, an OR gate is entered in column 2n-1).

$[M_i]$	$s_{2n-1}^{(i)}$	$s_{2n-2}^{(i)}$	......	$s_k^{(i)}$	.....	$s_1^{(i)}$	$s_0^{(i)}$
	$f_{2n-1}^{(i)}$	$f_{2n-2}^{(i)}$	......	$f_k^{(i)}$	.....	$f_1^{(i)}$	$f_0^{(i)}$
	$h_{2n-1}^{(i)}$	$h_{2n-2}^{(i)}$	......	$h_k^{(i)}$	.....	$h_1^{(i)}$	$h_0^{(i)}$
$[M_{i+1}]$	$s_{2n-1}^{(i+1)}$	$s_{2n-2}^{(i+1)}$	......	$s_k^{(i+1)}$	.....	$s_1^{(i+1)}$	$s_0^{(i+1)}$

It is evident that the following relationships apply:

1. $s_r^{(i)} \geq 3 \cdot f_r^{(i)} + 2 \cdot h_r^{(i)} \quad (0 \leq i \leq E-1)$,

since one fulladder comprises three line elements and one halfadder comprises two line elements of the same column.

2. $\quad s_r^{(i+1)} = s_r^{(i)} - 2f_r^{(i)} - h_r^{(i)} + f_{r-1}^{(i)} + h_{r-1}^{(i)}$, $(0 \leq i \leq E-1)$

since only the sum bits of fulladders and halfadders remain in column r (carries are transferred to column r+1), and the carries from column r-1 are taken over to column r.

3. (Calculation of M_{E+1} from the two-line matrix M_E with a carry ripple adder):

$$f_r^{(E)} = \begin{cases} 0 & r \leq m \vee s_r^{(E)} = 1 \\ 1 & \text{else} \end{cases} \qquad h_r^{(E)} = \begin{cases} 0 & r < m \vee s_k^{(E)} = 2 \\ 1 & \text{else} \end{cases}$$

where $m = \min \{i | s_i^{(E)} = 2\}$.

In order to achieve a fast reduction of M_0 to the two line matrix M_E or to the final matrix M_{E+1}, as many fulladders as possible shall be applied, since only these modules reduce the number of matrix elements.

For total effort G, we have:

$$G = n_{(3,2)} \cdot \kappa_{(3,2)} + n_{(2,2)} \cdot \kappa_{(2,2)} + n_{OR} \cdot \kappa_{OR},$$

where n_A and κ_A denote the number and costs of a type A logical module.

$$n_{(3,2)} = \sum_{i=0}^{E} \sum_{r=0}^{2n-1} f_r^{(i)} \quad ; \quad n_{(2,2)} = \sum_{i=0}^{E} \sum_{r=0}^{2n-1} h_r^{(i)}.$$

If the final addition is not included in the cost analysis, the first sum runs only to E-1 in each case.

3.4.3 Multiplication by serial, unclocked CSA addition

The two methods discussed can be demonstrated as follows (fig. 3.3); z_i denotes the i-th line of matrix M_0:

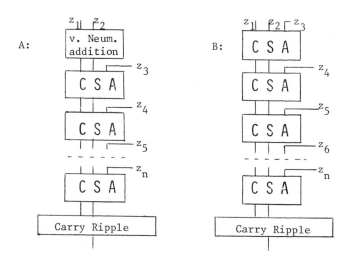

Fig. 3.3

Tables 3.3 and 3.4 show which modules are used at which positions in the CSA multiplication (F denotes a fulladder, H a halfadder).

Table 3.3 (CSA multiplication, method A)

column matrix	2n-1	2n-2	2n-3	2n-4	....	n+2	n+1	n	n-1	n-2	n-3	...	3	2	1	0
$[M_0]$	0	1	2	3	4...	n-3	n-2	n-1	n	n-1	n-2	..4	3	2	1	
								H	H	H			H	H	H	
$[M_1]$	0	1	2	3	4...	n-3	n-2	n	n	n-1	n-2	..4	3	1	1	
							F	F	F	F	..F	F				
$[M_2]$	0	1	2	3	4...	n-3	n-1	n-1	n-1	n-2	n-3	..3	1	1	1	

$[M_{n-2}]$	0	1	3	3	3...				3	1		1	1..	1	1	1
		F	F	F	....			F	F							
$[M_{n-1}]$	0	2	2	2	2...			2	1	1		....		1	1	1
		F	.	.	. ...		F	F	H							
$[M_n]$	1	1	1	....										1	1	1

Table 3.4 (CSA multiplication, method B)

matrix \ column	2n-1	2n-2	2n-3	....	n+2	n+1	n	n-1	n-2	n-3..3	2	1	0
$[M_0]$	0	1	2	3	4...n-3	n-2	n-1	n	n-1	n-2..4	3	2	1
							H	F	....			F	F H
$[M_1]$	0	1	2	3	4...n-3	n-1	n-1	n-1	n-2	n-3..3	2	1	1
						F	F	F	....				F H
$[M_2]$	0	1	2	3...n-4	n-2	n-2	n-2	n-2	n-3	n-4..2	1	1	1

$[M_{n-3}]$	0	1	3	3	3....3	3	3	3	2	1 1	....		1
			F	....	F	F	F	F	H				
$[M_{n-2}]$	0	2	2	2	2..............			2	1	1 1	....		1
		F	F	....			F	F	H				
$[M_{n-1}]$	1	1	1..										1

From tables 3.3 and 3.4 we immediately have for the total effort:

Method A: $n_{(3,2)} = (n-2) \cdot (n-1) + n-2 = n^2 - 2n$; $n_{(2,2)} = n$;

Method B: $n_{(3,2)} = (n-2) \cdot (n-1) - 1 + n-1 = n^2 - 2n$; $n_{(2,2)} = n$;

hence the same value in both cases. Both methods are optimum regarding effort (see 3.4.6).

3.4.4 Parallel multiplication, Wallace ([Wal] [Hal])

In this method, three adjacent lines of a matrix M_i are combined in a CSA adder respectively (see fig. 2.11). This interconnection is not determined uniquely, as is shown by the following two forms of a (8 x 8)-multiplier:

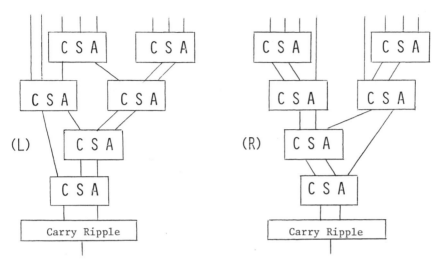

Fig. 3.4

However, the number of E steps of the adder-tree and the number of the CSA adders necessary are determined uniquely:

<u>Proposition 3.3.</u> *a. For a $(n \times n)$-multiplier according to Wallace we need exactly $n-2$ CSA adders for reducing M_0 to a two line matrix M_E.*

b. For the number of steps we estimate as follows:

$$\frac{1}{\log_2 1{,}5} \cdot [\log_2(n) + \log_2 \frac{9}{16}] \leq E \leq \frac{1}{\log_2 1{,}5} \cdot [\log_2(n-2)-1] + 2 \;;$$

i.e. $E \approx 1{,}72 \cdot \log_2(n)$ *when* $n \geq 4$.

Proof is elementary.

For all applications, 12 steps are sufficient at most, as is shown in table 3.5 .

Table 3.5 (number of E steps of the Wallace-tree for length of the addend n)

E	n	E	n	E	n	E	n
0	1,2	4	7,8,9	7	$20 \leq n \leq 28$	10	$64 \leq n \leq 94$
1	3	5	$10 \leq n \leq 13$	8	$29 \leq n \leq 42$	11	$95 \leq n \leq 141$
2	4	6	$14 \leq n \leq 19$	9	$43 \leq n \leq 63$	12	$142 \leq n \leq 211$
3	5,6						

<u>Effort.</u> In addition to (3,2)- and (2,2)-counters, OR gates with two inputs are also necessary for obtaining the CSA. The OR gates are necessary only in column 2n-1.

When carrying out final addition by means of a carry ripple adder, the following relationships generally apply (see 3.4.6):

$$n_{(3,2)} + n_{OR} = n^2 - 2n \; ; \; n_{(2,2)} \geq n \; .$$

The effort depends on the form chosen. For (8 x 8)-Wallace multiplications, we have for example (fig. 3.4):

	$n_{(3,2)}$	n_{OR}	$n_{(2,2)}$
Form (L)	48	0	15
Form (R)	45	3	23

In both cases $n_{(2,2)}$ is much larger than the lower limit $n_{(2,2)} = 8$.

<u>Cycle time.</u> The "Wallace-tree" can be organised in such a way that a new bit of the product can be calculated in each i ($1 \leq i \leq E$) step. Addition of both M_E lines then refers to two addends of length

$$2n-1 - (E+1) \approx 2n - \frac{\log_2 n}{\log_2 1,5} - 2 \quad \text{(see [Hal])} \; .$$

Assuming two logical steps for the CSA addition and employing a
conditional sum adder (cycle time = $2 \cdot \lceil \log_2 m \rceil + 4$ for addends
of length m) for final addition, for the total cycle time of the
Wallace multiplier we obtain:

$$\tau_1 = 2 \cdot E + 2 \cdot \lceil \log_2(2n-E-2) \rceil + 4 \approx 6 \cdot \log_2 n$$

(where $16 \leq n \leq 128$).

Multiplier coding

The Wallace-tree can be reduced by two steps at best by coding the
multiplier in groups of 2+1 bits each. The resulting time saving
is lost because of the cycle time necessary for coding. Multiplier
coding with parallel multiplications is practicable only when applying pipeline techniques, i.e. if several parts of the multiplication
overlap. The computing time for coding, entering into the overall
cycle time for multiplication is thus reduced to a fraction of the
original value. A detailed discussion of the pipelining scheme can
be found in 3.6 .

3.4.5 Parallel multiplication, Dadda scheme [Da1], [Da2]

In contrast to Wallace multiplication, where 3 <u>lines</u> of one matrix are
combined respectively into a carry save adder, parallel multiplication
according to Dadda results in the sums of the <u>columns</u> occupying a
specific area.

The Dadda multiplier requires as many steps as the Wallace multiplication; it employs exclusively (3,2)- and (2,2)-counters. The following
describes one version of this method and we can demonstrate that it is
the optimum as regards effort.

Let max(E) be the largest number in the line of a multiplication
matrix M_0, which can be reduced with a Wallace-tree of E steps to a
matrix M_E of 2 lines, for example max(9) = 63 (see table 3.5) .

Multiplication procedure according to Dadda

1. Determine $E=E(n)$ with $\max(E-1) < n \leq \max(E)$.

2. Reduce the maximum column sum M_{i+1} to $\max(E-i-1)$ ($i=0,\ldots,$ E-1) with the minimum effort and further determine a new product bit at every step of the procedure. This rule can be described clearly as follows (indication of the number of full- and half-adders employed in column r of M_i):

$$f_r^{(i)} := \begin{cases} \lfloor \frac{s_r^{(i)}}{3} \rfloor & r = 0,\ldots,n \\ f_{r-1}^{(i)} & r > n \wedge s_r^{(i)} = s_{r-1}^{(i)} \\ \max(f_{r-1}^{(i)} + h_{r-1}^{(i)} - 1, 0) & r > n \wedge s_r^{(i)} \neq s_{r-1}^{(i)} \end{cases}$$

$$h_r^{(i)} := \begin{cases} 1 & r = i+1 \vee (s_r^{(i)} + f_{r-1}^{(i)} + h_{r-1}^{(i)} - 2f_r^{(i)} > \max(E-i-1)) \\ 0 & \text{otherwise.} \end{cases}$$

Remark. Following transformation from M_i to M_{i+1}, $s_r^{(i+1)}$ must not exceed the maximum value which can be reduced in E-i-1 steps. If $s_r^{(i+1)} = s_r^{(i)} + f_{r-1}^{(i)} + h_{r-1}^{(i)} - 2f_r^{(i)}$ exceeds the value max (E-i-1) in step i after application of $f_r^{(i)}$ fulladders, an additional half-adder must be employed in column r (see following examples). The formulae for $f_r^{(i)}$ and $h_r^{(i)}$ show that only as many fulladders are applied as regards the left half of the multiplication (columns r=n+1,...,2n-1) as are absolutely necessary for reducing $s_r^{(i+1)}$ to a value $\leq \max(E-i-1)$.

3. Carry ripple addition of both addends of M_E in step E+1.

The following characteristics of this multiplication procedure can be demonstrated readily by induction:

Lemma 3.4. a. If $h_r^{(i)} = 1 \wedge r \neq i+1$, then:

$$(s_{2n-1}^{(i+1)}, \ldots, s_r^{(i+1)}) = (0,1,2,\ldots,m-2,m,m,\ldots,m),$$

whence $m = max(E-i-1)$,

b. In columns $1,\ldots,n$, one halfadder is employed at most in each case, and none is used in the remaining columns;

i.e. $h_r^{(i)} = 0$, if $r \notin \{1,\ldots,n\}$;

$h_r^{(i)} = 1 \Rightarrow h_r^{(j)} = 0$ for $j \neq i$.

<u>Examples</u>

1. n=9, i.e. E=4 ; (halfadders are circled, fulladders are not)

column matrix	17								10	9	8	7	6	5	4	3	2	1	0	
$[M_0]$	0	1	2	3	4	5	6	7	8	9	8	7	6	5	4	3	2	1		
								1	2	2	3	2	2	2	1	1	1			
					(*)⟶①								①							
$[M_1]$	0	1	2	3	4	6	6	6	6	5	6	5	3	4	3	2	1	1		
						1	2	2	2	2	1	2	1	1	1					
									①				①							
$[M_2]$	0	1	2	4	4	4	4	4	4	3	4	2	3	2	1	1	1			
				1	1	1	1	1	1	1	1	1①								
$[M_3]$	0	1	3	3	3	3	3	3	3	2	2	3	2	1	1	1				
		1	1	1	1	1	1	1	①①1①											
$[M_4] = [M_E]$	0	2	2	2	2	2	2	2	2	2	2	2	1	1	1	1				
		1	1	1	1	1	1	1	1	1	1①									
$[M_5] = [M_{E+1}]$	1	1	1	1	1	1	1	1	1	1	1	1	1	1	1	1	1			

$\left.\begin{array}{c}\\\\\end{array}\right\}$ carry ripple addition

On transferring from M_0 to M_1, the maximum sum of column M_1 is reduced to a value which does not exceed max(4-0-1) = 6. Without the halfadder marked (*), we would have $s_9^{(1)} = 7$; these 7 column elements could no longer be reduced to two elements in the remaining 3 steps.

2. n=16 ⇒ E=6 .

```
[M₀] |0 1 2 3 4 5 6 7 8 9 10 11 12 13 14 15 16 15 14 13 12 11 10 9 8 7 6 5 4 3 2 1
                         1  2  3  4  5  5  5  4  4  4  3  3 3 2 2 2 1 1 1①
[M₁] |0 1 2 3 4 5 6 7 8 9 11 11 11 11 11 10 11  9 10  9  7  8 7 5 6 5 3 4 3 2 1 1
                         1  2  3  3  3  3  3  3  3  3  3 2 2 2 1 2 1 1 1 1①
[M₂] |0 1 2 3 4 5 6 8 8 8  8  8  8  8  8  7  8  6  7  5  5  6 4 5 3 4 2 3 2 1 1 1
                      1  2  2  2  2  2  2  2  2  2  2  2  1  1 2 1 1 1 1 1①
[M₃] |0 1 2 3 4 6 6 6 6 6  6  6  6  6  5  6  4  4  4  5  3  3 4 2 2 3 2 1 1 1 1
                 1  2  2  2  2  2  2  2  2  2  ¹②  2  1  1  1  1  1 1 1 1①
[M₄] |0 1 2 4 4 4 4 4 4 4  4  4  4  4  4  3  3  3  3  4  2  2 2 2 3 2 1 1 1 1 1
              1  1  1  1  1  1  1  1  1  1  1  1  1  1  1  1                1①
[M₅] |0 1 3 3 3 3 3 3 3 3  3  3  3  3  3  2  2  2  2  2  2  2 2 3 2 1 1 1 1 1 1
              1  1  1  1  1  1  1  1  1  1  1  ① ① ① ① ① ① ① ① 1①
[M₆] |0 2 2 2 2 2 2 2 2 2  2  2  2  2  2  2  2  2  2  2  2  2 2 2 1 1 1 1 1 1 1
              1  1  1  1  1  1  1  1  1  1  1  1  1  1  1  1  1  1  1 1①
[M₇] |1 1 1 1 1 1 1 1 1 1  1  1  1  1  1  1  1  1  1  1  1  1 1 1 1 1 1 1 1 1 1
```

3.4.6 Investigation of effort

We will demonstrate that the version of the Dadda multiplier, given by us, is optimum with regard to the total cycle time and the number of modules employed.

<u>Proposition 3.5.</u> Let $\max(E-1) < n \leq \max(E)$, $(n \geq 2)$. Then:

a. Each multiplication procedure requires at least E steps for reducing M_0 to a two line matrix M_E.
b. The final addition extends over 2 addends of length $\geq 2n-1-(E+1)$.
c. The Wallace and Dadda methods have optimum cycle times (minimum number of steps, minimum length of final addition).
d. For all forms beyond the module system quoted (independent from choice of final adder), we have:

$$n_{(3,2)} + n_{OR} = n^2 - 2n \; ; \; n_{(2,2)} \geq n \, .$$

If $n_{(2,2)} = n$, one HA is applied in columns $1, 2, \ldots, n$ respectively. OR gates are not employed in this case.

e. For the version of Dadda multiplier quoted and for the CSA multiplier (see 3.4.3), we obtain (if final addition is carried out with a carry ripple adder):

$$n_{(3,2)} = n^2 - 2n, \; n_{OR} = 0, \; n_{(2,2)} = n \, .$$

In general, the Wallace multiplication yields:

$$n_{OR} > 0 \; ; \; n_{(2,2)} = n + \varepsilon \quad \text{with } \varepsilon > 0 \; ;$$

here, ε depends on the form, and may be larger than n (see the two forms in 3.4.4; there we have: $\varepsilon = n-1$ or $\varepsilon = 2n-1$).

<u>Proof.</u> All statements except the last two follow from earlier considerations.

d. Let $|[M_i]| := \sum_{r=0}^{2n-1} s_r^{(i)}$ be the element number of the matrix $[M_i]$.

Each fulladder and each OR gate with two inputs reduces the element number by 1, while a halfadder does not provide any reduction.

Since $|[M_0]| = n^2$, $|[M_{E+1}]| = 2n$, it follows that
$n_{(3,2)} + n_{OR} = n^2 - 2n$.

We will demonstrate by induction that at least n halfadders are required.

```
           0  1  2  1      [M_0]  ⎫
           ─────────              ⎪
              ①                   ⎪  Dadda multi-
           0  2  1  1      [M_1]  ⎬  plication
           ─────────              ⎪  scheme for
              ①                   ⎪  n = 2
           1  1  1  1      [M_2]  ⎭
```

Apart from 2 halfadders, no other modules are used for n = 2 (see diagram above). In order to convert from n to n+1, we split up a (n+1)-line matrix, obtained from

$$[M_0] = [0,1,2,3,\ldots,n,n+1,n,\ldots,3,2,1]$$

in the following manner:

$$[M_0] = [M_0'] + [M_0''] = [0,0,1,2,\ldots,n-1,n,n-1,\ldots,2,1,0]$$
$$+ [0,1,1,\ldots\ldots\ldots\ldots\ldots\ldots,1,1,1].$$

By induction, the n-line matrix M_0' can be reduced by means of $n^2 - 2n$ fulladders and n halfadders (in columns $1,\ldots,n$ of M_0', i.e. in columns $2,\ldots,n+1$ of M_0) (column 0 of a matrix signifies throughout the column on the far right with a positive element number). The result of this reduction is the matrix

$$[M_0^*] = [M_E'] + [M_0''] = [0,1,1,\ldots\ldots\ldots,1,1,0]$$
$$+ [0,1,1,\ldots\ldots\ldots,1,1,1]$$
$$= [0,2,2,\ldots\ldots\ldots,2,2,1].$$

Finally, M_0^* is reduced by means of a halfadder (in column 1 of M_0) and a fulladder (in column $2,3,\ldots,2n$) respectively. Hence, for the total effort for reduction of M_0 (i.e. for multiplication of two (n+1)-digit numbers), we have:

$$n_{(2,2)} = n + 1; \quad n_{(3,2)} = n^2 - 2n + 2n - 1 = (n+1)^2 - 2(n+1) .$$

It can be readily demonstrated that the cycle time for multiplication (number of steps) can vary with other types of reduction (i.e. other splittings of M_0), but the minimum number of fulladders and halfadders remains unchanged. The proposition is thus proved.

e. In regard to the versions of Dadda multiplier quoted, one halfadder at most is used in columns $1,2,\ldots,n$ (see lemma 3.4). Since a total of n halfadders is necessary, this method is the optimum regarding effort.

3.4.7 Multiplication with factors of various lengths

Let $MD = [MD_{k-1},\ldots,MD_0]$; $MQ = [MQ_{n-1},\ldots,MQ_0]$; $MD_i, MQ_i \in \{0,1\}$.

$$\Rightarrow [M_0] = \begin{cases} [1,2,\ldots,k,k,\ldots,k,k-1,\ldots,1] & \text{where } k \leq n \\ [1,2,\ldots,n,n,\ldots,n,n-1,\ldots,1] & \text{where } k \geq n \end{cases}$$

$$\longleftarrow n + k - 1 \longrightarrow$$

For multiplication matrices of this form we have:

Proposition 3.6. *a. Matrix $[M_0]$ is independent from the sequence of factors. In order to reduce M_0 to a two line matrix, at least $E = E(n,k)$ steps are required, where*

$$max(E-1) < min(n,k) \leq max(E) .$$

The Dadda multiplication meets these requirements; with the definition of $f_r^{(i)}$ and $h_r^{(i)}$ (number of fulladders or halfadders in step i applied in column r, see 3.4.5), max(n,k) is written instead of n in each case.

b. *The Dadda multiplier is optimum regarding effort:*

$$n_{(3,2)} = n \cdot k - (n+k) \; ; \quad n_{(2,2)} = min(n,k) .$$

One halfadder in each case is applied in columns 1,2,......, min(n,k)-1 as well as in column max(n,k) of M_0.

c. *With the CSA multiplication and the Wallace method, the cycle time or effort depends on the sequence of the factors; this is not the case with the Dadda method.*

<u>CSA multiplication:</u>

$$n_{(3,2)} = n \cdot k - (n+k); \quad n_{(2,2)} = k .$$

<u>Wallace multiplication:</u>

$$n_{(3,2)} + n_{OR} = n \cdot k - (n+k); \quad n_{(2,2)} \geq k .$$

This proposition can be proved by simple generalisation of the considerations of 3.4.5 . We will demonstrate the Dadda multiplication procedure, taking as example a 15 x 13 multiplier:

[M₀]	0 1 2 3 4 5 6 7 8 9 10 11 12 13 13 13 12 11 10 9 8 7 6 5 4 3 2 1
	1 2 3 4 4 4 4 4 3 3 3 2 2 2 1 1 1 ①
[M₁]	0 1 2 3 4 5 6 7 9 9 9 9 9 9 9 9 7 8 7 5 6 5 3 4 3 2 1 1
	1 2 3 3 3 3 $\overset{2}{①}$ 3 3 3 2 2 2 1 2 1 1 1 1 ①
[M₂]	0 1 2 3 4 6 6 6 6 6 6 6 6 6 6 5 5 6 4 5 3 4 2 3 2 1 1 1
	1 2 2 2 2 2 2 2 2 2 $\overset{1}{①}\overset{1}{①}$ 2 1 1 1 1 1 ①
[M₃]	0 1 2 4 4 4 4 4 4 4 4 4 4 4 4 4 4 3 3 4 2 2 3 2 1 1 1 1
	1 1 1 1 1 1 1 1 1 1 1 1 1 1 1 1 1 ①
[M₄]	0 1 3 3 3 3 3 3 3 3 3 3 3 3 3 3 3 2 2 2 2 3 2 1 1 1 1 1
	1 1 1 1 1 1 1 1 1 1 1 1 1 1 1 ①①①①1①
[M₅]	0 2 2 2 2 2 2 2 2 2 2 2 2 2 2 2 2 2 2 2 2 2 1 1 1 1 1 1	⎤ carry
	1 1 1 1 1 1 1 1 1 1 1 1 1 1 1 1 1 1 1 ①	⎬ ripple
[M₆]	1 1 1 1 1 1 1 1 1 1 1 1 1 1 1 1 1 1 1 1 1 1 1 1 1 1 1 1	⎦ addition

3.4.8 Multiplication with base 2^h

3.4.8.1 General principle

The number of lines of the basic matrix M_0 for multiplication can be lowered by coding the multiplier and multiplicand on a higher base. In particular, all codings with base $d = 2^h$ ($h \in \mathbb{N}$) are suitable for treatment.

Using this base, we obtain the following new representation for multiplicand and multiplier:

$$MD^{(h)} = [MD^{(h)}_{k_h-1}, \ldots, MD^{(h)}_0], \quad \text{where } 0 \le MD^{(h)}_i \le 2^h - 1$$

$$\text{and } k_h := \lceil \tfrac{k}{h} \rceil ;$$

$$MQ^{(h)} = [MQ^{(h)}_{n_h-1}, \ldots, MQ^{(h)}_0], \quad \text{where } 0 \le MQ^{(h)}_i \le 2^h - 1$$

$$\text{and } n_h := \lceil \tfrac{n}{h} \rceil .$$

Restricting the case to non-negative numbers, we obtain:

$$w(MD^{(h)}) = \sum_{i=0}^{k_h-1} MD^{(h)}_i \cdot 2^{hi}, \quad w(MQ^{(h)}) = \sum_{i=0}^{n_h-1} MQ^{(h)}_i \cdot 2^{hi} .$$

The multiplication matrix M_0 consists of $k_h \times n_h$ elements:

$$M_0^{(h)} = \begin{bmatrix} 0 & & m^{(h)}_{k_h-1,0} & \cdots\cdots\cdots & m^{(h)}_{0,0} \\ & \cdot & & \cdot & \\ & \cdot & & \cdot & \\ m^{(h)}_{k_h-1,n_h-1} & \cdots\cdots\cdots & m^{(h)}_{0,n_h-1} & 0 & \end{bmatrix} \Bigg\} n_h$$

$$\underbrace{\hspace{6cm}}_{k_h}$$

where $m^{(h)}_{i,j} := MD^{(h)}_i \cdot MQ^{(h)}_j \;;\quad 0 \le m^{(h)}_{i,j} \le (2^h - 1)^2 .$

The binary representation of an element $m^{(h)}_{i,j}$ has the length

$$u_h := \lceil \log_2((2^h-1)^2 + 1) \rceil = \begin{cases} 1 & h = 1 \\ 2h & h > 1 \end{cases} \text{ bits .}$$

For a binary coding of one line, we require at least 2 lines for $h > 1$; this number of lines is always adequate.

<u>Example.</u> (Representation by A+S; no indication of sign in M_0).

Let $h = 2$, i.e. $0 \le m^{(2)}_{i,j} \le 9$;

$$MD^{(2)} = [2,3,1,0,3] \; ; \; MQ^{(2)} = [3,1,2] \; ; \text{ i.e. } k_2=5, \, n_2=3.$$

$$M_0^{(2)} = \begin{bmatrix} 4 & 6 & 2 & 0 & 6 \\ 2 & 3 & 1 & 0 & 3 \\ 6 & 9 & 3 & 0 & 9 \end{bmatrix}$$

$$M_0 = \begin{bmatrix} 0 & 0 & 1 & 0 & 1 & 0 & 0 & 0 & 1 & 0 \\ & & 0 & 1 & 0 & 1 & 0 & 0 & 0 & 0 & 0 & 1 \\ & & 1 & 0 & 1 & 1 & 0 & 1 & 0 & 0 & 1 & 1 \\ & & 0 & 0 & 0 & 0 & 0 & 0 & 0 & 0 & 0 & 0 \\ & & & & 1 & 0 & 0 & 1 & 1 & 1 & 0 & 0 & 0 & 1 \\ & & & & 0 & 1 & 1 & 0 & 0 & 0 & 0 & 0 & 1 & 0 \end{bmatrix}$$

Each element of the binary coding M_0 of matrix $M_0^{(2)}$ must be coded by $u_2=4$ bits. M_0 consists of $s_2 = 2 \cdot \min(n_2,k_2) = 6$ lines.

For the maximum binary column sum of M_0, we have generally:

$$s_h \leq 2 \cdot \min(n_h, k_h) \, .$$

The effort for determining $m_{i,j}^{(h)}$ increases with increasing h; hence it is inappropriate to use high values for h.

3.4.8.2 Multiplication with base 4, using (4,2)-counters

(Ferrari, Stefanelli, 1969) [Fe3]

Multiplication with base 4 can be accelerated with the (4,2)-counters discussed in 2.1:

Let $MD = [MD_{k-1},\ldots,MD_0]$; $MQ_{n-1} = [MQ_{n-1},\ldots,MQ_0]$;

$$m_{i,j}^{(2)} := MD_i^{(2)} \cdot MQ_j^{(2)} \quad (0 \leq m_{i,j}^{(2)} \leq 9) \, .$$

Since all $m_{i,j}^{(2)}$ are coded with 4 binary figures, but since only 10 of 16 representable numbers can occur, we can specify the form of multiplication matrix M_0 which can be used for quicker reduction to a two line matrix (see also example 3.4.8.1).

Lemma 3.7. *Each 4 successive elements of a column of the binary multiplication matrix M_0 contains at least one zero.*

Proof of this lemma is provided in [Fe3].

The multiplier based on this lemma also uses (4,2)-counters in addition to FA and HA in the first step (see 2.1). After the first step, the number of lines is reduced to $\lceil\frac{n}{2}\rceil$ or $\lceil\frac{n-1}{2}\rceil$ (contrary to $\approx \frac{2}{3} \times n$ by the usual method). The number of steps is therefore reduced generally by 1. However, the gain in cycle time is again lost due to the greater effort (for costs and time) for coding the basic matrix, and due to the increased costs for the (4,2)-counters.

3.5 Arithmetic circuits [Me2]

3.5.1 Definitions

The use of the (4,2)-counter in the method described under 3.4.8.2 raises the question as to what extent more general modules would be suitable for parallel multiplication or for other arithmetic operations. The development of integrated circuits renders generally possible the application of such modules, which would be out of the question for financial reasons with today's techniques. The following provides a general description of such modules (called arithmetic circuits):

Definition 3.3. *a. A circuit is called an <u>arithmetic circuit</u>, if there is a partition*

$$E_0 = \{e_0^{(1)}, \ldots, e_0^{(m_0)}\}, \ldots, E_p = \{e_p^{(1)}, \ldots, e_p^{(m_0)}\}$$

of the set of circuit inputs and a partition

$$A_0 = \{a_0^{(1)}, \ldots, a_0^{(n_0)}\}, \ldots, A_p = \{a_p^{(1)}, \ldots, a_p^{(n_p)}\}$$

of the set of outputs with:

$$\sum_{i=0}^{p} \sum_{j=1}^{m_i} e_i^{(j)} \cdot 2^i = \sum_{i=0}^{p} \sum_{j=1}^{n_i} a_i^{(j)} \cdot 2^i$$

for all permissible combinations of input variables $e_i^{(j)}$.

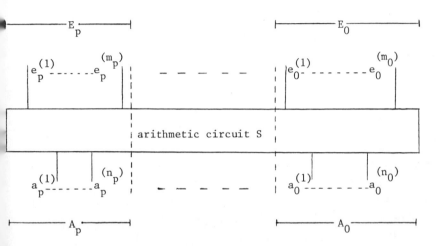

Fig. 3.5

b. An arithmetic circuit is triangular, if

$$|A_i| \leq 1 \quad (i = 0, \ldots, p).$$

If the circuit is invariant as regards permutations within the quantities $E_0, \ldots, E_p$ and $A_0, \ldots, A_p$ (as is the case with almost all applications), it is simpler to describe the circuit by the number of inputs and outputs of the individual classes. In the following we will examine such circuits only. Furthermore we can characterize a circuit by finding how much it reduces or increases the number of outputs as compared with the respective number of inputs:

<u>Definition 3.4.</u> *a*. $[m_p,\ldots,m_0]$ or $[n_p,\ldots,n_0]$ *indicate input/output sequence*; m_i and n_i denote the <u>number</u> of inputs/ outputs of the individual classes of the arithmetic circuit.

b. $<m_p - n_p, \ldots, m_0 - n_0>$ indicates <u>characteristic sequence (CS)</u>;

c. A triangular circuit with CS = $<c_p,\ldots,c_0>$ is called <u>compact</u>, if: 1. $c_i \geq -1$ for all i;
2. $c_p = -1$;
3. $c_j = -1 \Rightarrow c_{j+1} = \ldots = c_p = -1$.

In general, the characteristic sequence does not unambiguously determine the structure of a circuit.

<u>Examples.</u>

a. <u>Multiplication of n-digit numbers</u>

Input sequence: $[0,1,2,\ldots,n-1, n, n-1,\ldots,2,1]$
Output sequence: $[1,1,1,\ldots\ldots\ldots\ldots\ldots\ldots,1,1]$ (see 3.4.2) .
Char. sequence: $<-1,0,1,\ldots,n-2,n-1,n-2,\ldots,1,0>$.

b. <u>Fulladder</u>

Input sequence: $[0,3]$
Output sequence: $[1,1]$
Char. sequence: $<-1,2>$

c. <u>Halfadder</u>

Input sequence: $[0,2]$
Output sequence: $[1,1]$
Char. sequence: $<-1,1>$

d. <u>(31,5)-counter</u>

Input sequence: $[0, 0, 0, 0, 31]$
Output sequence: $[1, 1, 1, 1, 1]$
Char. sequence: $<-1,-1,-1,-1, 30>$.

All cases involve compact circuits.

Remark. AND and OR gates with at least two inputs are not arithmetic circuits, as can be seen readily from the definition 3.3; (the sum of the inputs differs from the sum of the outputs for at least one input combination).

3.5.2 Interconnection of arithmetic circuits

Assume a quantity $\{S_1, \ldots, S_k, \ldots\}$ of arithmetic circuits. On shifting one circuit to the left by r digits (multiplication by 2^r) we obtain a new arithmetic circuit (of order r), which comprises the same elements. In this way, an <u>absolute weight</u> is associated with each input or output line. Fig. 3.6 shows an example.

Basically, there are two options for interconnecting arithmetic circuits:

A. <u>Series connection</u> $S_1 \circ S_2$

1. Each S_1 output can be connected at most with one S_2 input.
2. Interconnection of two lines is permissible only if they have the same absolute weight.

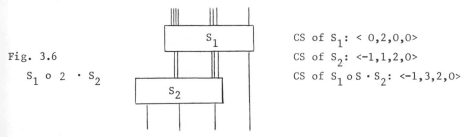

Fig. 3.6
$S_1 \circ 2 \cdot S_2$

CS of S_1: $<0,2,0,0>$
CS of S_2: $<-1,1,2,0>$
CS of $S_1 \circ S \cdot S_2$: $<-1,3,2,0>$

B. <u>Parallel connection</u> $S_1 \otimes S_2$

Let $\alpha_j^{(i)}$ be the number of inputs of S_i with absolute weight 2^j;
Let β_j be the number of available input lines with absolute weight 2^j.
Parallel connection of S_1 and S_2 is possible, if:

$$\alpha_j^{(1)} + \alpha_j^{(2)} \leq \beta_j \quad \text{for} \quad j = 0,1,2,\ldots \;.$$

Remark. On interconnecting arithmetic circuits by the operations "o" or "$\otimes$", we again obtain arithmetic circuits (see fig. 3.7).

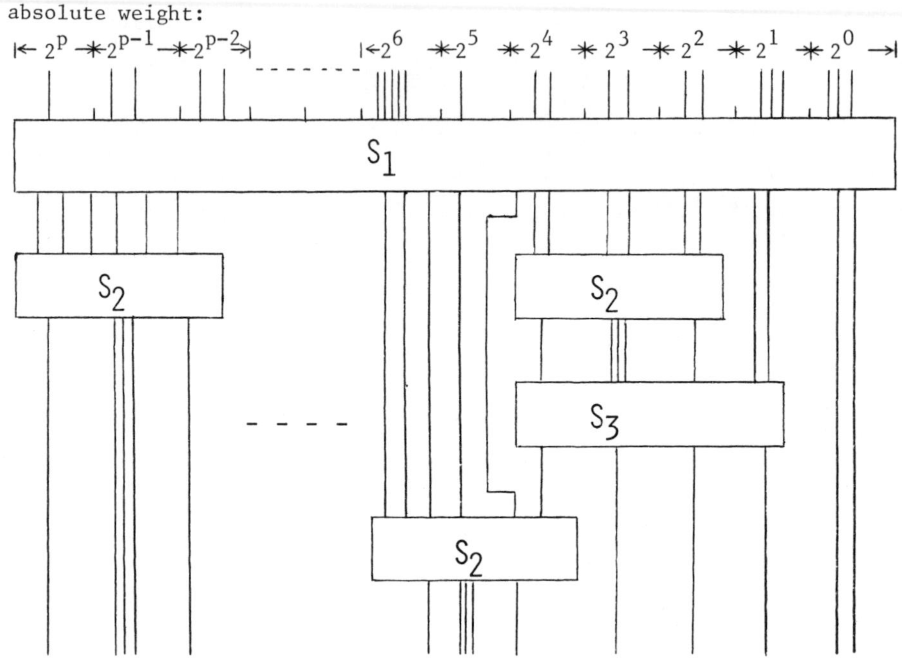

absolute weight:

$|\leftarrow 2^p \ast 2^{p-1} \ast 2^{p-2} \rightarrow| \quad \cdots \quad |\leftarrow 2^6 \ast 2^5 \ast 2^4 \ast 2^3 \ast 2^2 \ast 2^1 \ast 2^0 \rightarrow|$

Fig. 3.7

Circuit S_2 is inserted at three positions (order 2, 4 or p-2).

$$S = S_1 o (2^{p-2} \cdot S_2 \otimes 2^2 \cdot S_2)$$
$$o \ 2 \cdot S_3 \ o \ 2^4 \cdot S_2$$

3.5.3 Reduction problem

Let the input sequence be

$$E = [\ldots, 0, e_m, e_{m-1}, \ldots, e_0, 0, \ldots] \quad \text{(i.e. } e_i = 0 \text{ where } i \notin [0:m]\text{)}.$$

The non-vanishing section of E may be shifted to the left or right by multiplication or division, without any substantial changes.

We are seeking a system of arithmetic elementary circuits as simple as possible, by which E can be transformed to an r-reduced output sequence A_r.
(A sequence is r-reduced, if $0 \leq u \leq r$ applies for every element u of the sequence).

Above all, the 2-reduced output sequences (for example the two-line final matrix with parallel multiplication) and the 1-reduced output sequences (for example the final result of an arithmetic operation) are of interest in practice.

Since solution of the reduction problem is too difficult in the general form stated, we restrict ourselves to triangular or even compact modules. Nearly all reasonable modules meet this condition. All others can be constructed from triangular elements.

<u>Remark.</u> A circuit for the r-reduction of $E = [e_m, \ldots, e_0]$ is also an r-reduction for all $E' = [e'_m, \ldots, e'_0]$ with $0 \leq e'_i \leq e_i$ (i.e. for $E' \leq E$). An r-reduction of E which is optimal regarding costs and cycle time is generally not optimal for $E' \leq E$.

3.5.4 Reduction with triangular elementary circuits

The reduction of an input sequence

$$E = [e_m, e_{m-1}, \ldots, e_0] \quad (e_i := 0 \text{ if } i \notin [0:m])$$

by a triangular circuit S (given by its characteristic sequence $<c_p, \ldots, c_0>$) can be described as follows:

$$E = [e_m\ e_{m-1}\ldots e_{r+p+1}\ e_{r+p}\ e_{r+p-1}\ldots e_{r+1}\ e_r\ e_{r-1}\ldots e_0]$$
$$<c_p\ c_{p-1}\ \ldots\ c_1\ c_0>$$

$$E^* = [e_m^*\ e_{m-1}^*\ldots e_{r+p+1}^*\ e_{r+p}^*\ e_{r+p-1}^*\ldots e_{r+1}^*\ e_r^*\ e_{r-1}^*\ldots e_0^*]$$

where

$$e_i^* = \begin{cases} e_i & \text{where } i \notin [r:r+p] \\ \max\{e_i - c_{i-r};\ 1\} & \text{otherwise} \end{cases}$$

<u>Reason.</u> If $e_i > c_{i-r}$ (i $\in$ [r:r+p]), then the number of outputs is reduced to $e_i - c_{i-r}$ by S. However, if $e_i \le c_{i-r}$, we must increase the number of E inputs by lines constantly resulting in zero value until, after reduction, one output remains. The following secondary lemma shows that a reduction to 0 outputs is impossible with regard to triangular circuits:

<u>Lemma 3.8.</u> *For triangular circuits with characteristic sequence* $<m_p - n_p, \ldots, m_0 - n_0>$ *we have:*

a. $m_k \ge 1 \Rightarrow n_k = 1$;

b. $n_k = 0 \Rightarrow n_k = m_k = 0$.

<u>Proof.</u> We will consider a special input combination in which precisely one of the m_k inputs of E_k leads with a 1, all other inputs of the circuit are occupied by zero.
Assuming that the circuit is triangular, it must have at least one output at the k position, i.e. n_k=1; this need not apply for non-triangular circuits.

<u>Remark.</u> One module S with CS = $<c_p, \ldots, c_0>$ can replace all modules S' with CS = $<c_p', \ldots, c_0'>$ ($c_i' \le c_i$). Instead of a halfadder (CS =

<-1,1>) we can use, for example, a fulladder (CS = <-1,2>).
The costs for S' are usually less than for S.

3.5.5 Reduction of the multiplication matrix with <-1,0,3,2>-elements

A <-1,0,3,2>-element is a compact arithmetic circuit which can be obtained by interconnecting 4 fulladders. The number of outputs is less by 4 than the number of inputs.

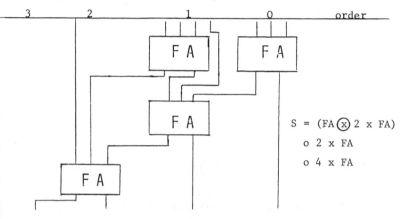

Fig. 3.8 (<-1,0,3,2>-element)

At least $\lceil \frac{1}{4} \cdot (n^2 - 2n) \rceil$ modules are necessary for the 1-reduction of a multiplication matrix for n-digit factors, using exclusively this circuit element. This number increases to $\lceil \frac{1}{4} \cdot (n^2 + a - 2n) \rceil$, if an additional input lines are used which are constantly occupied by zero.

The following reflections demonstrate that <-1,0,3,2>-elements are well suited to the 1- or the 2-reduction of a multiplication matrix. We will use the notation $<d \cdot c_p, d \cdot c_{p-1}, \ldots, d \cdot c_0>$, if the circuit element $<c_p, \ldots, c_0>$ is used d times at the same positioning. These d reductions can be carried out simultaneously.

3.5.5.1 1-reduction

Serial reduction from right to left involves minimal effort; it corresponds to the CSA multiplication (see 3.4.3). For n=9, the method operates as follows:

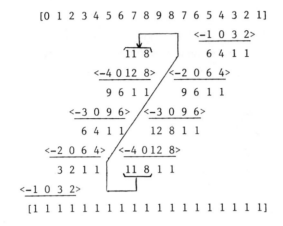

Multiplication can be accelerated by parallel performance of as many reduction steps as possible. In this way an algorithm is obtained corresponding to the multiplication procedure according to Dadda (3.4.5).

It can readily be demonstrated that the following applies to the serial 1-reduction as stated:

<u>Cycle time:</u> n-1 steps.

<u>Costs:</u> $\lceil \frac{1}{4} \cdot (n^2 - 1) \rceil$ modules of the <-1,0,3,2> type.

This is to be compared with the values of the CSA multiplication:

<u>Cycle time:</u> 2(n-2) steps;

<u>Costs:</u> $n^2 - n$ modules ($n^2 - 2n$ fulladders and n halfadders).

If the <-1,0,3,2>-element is obtained by means of 4 fulladders, the gain in cycle time (halving the number of steps) is lost, or again the reduction in number of modules to approximately ¼ as against the CSA multiplication. A two step form of the element with AND and OR gates consists of at least 179 AND gates (with 5 inputs on average) and of 4 OR gates (with 4, 48, 84 and 43 inputs [Me2]). With the modern state of the art even a form employing integrated circuits would be 10 times as costly as one fulladder. However, it is possible that the costs may in the near future vary to such an extent in favour of more generalized arithmetic modules that these circuit elements may become competitive.

3.5.5.2 2-reduction

A simpler 2-reduction can be obtained by placing the first <-1,0,3,2>-module only in the third column. We will demonstrate the procedure for n=9. It can be readily shown that the method can be applied generally.

```
            [0 1 2 3 4 5 6 7 8 9 8 7 6 5 4 3 2 1]
                                <-1 0 3 2>
                       10 7      7 5 1 1
                <-3 0 9 6>  /<-2 0 6 4>
                  7 5 1 1 /   10 7 1 1
             <-2 0 6 4>  /<-3 0 9 6>
                4 3 1 1 /   11 9 1 1
         <-1 0 3 2>  /<-4 012 8>
            1 1 1 1  | 10 7 1 1
            [1 1 1 1 1 1 1 1 1 1 1 1 1 1 1 2 1]
```

For reducing [0,1,2,...,n-1,n,n-1,...,3,2,1] to the output
sequence [1,1,1,............,1,1,1,2,1] , we require:

Cycle time: n-2 steps;

Costs: $\lceil \frac{1}{4} \cdot (n^2 - 2n) \rceil$ modules of the <-1,0,3,2> type.

This reduction is optimal with regard to effort. A disadvantage,
however, is the fact that final calculation of the product (1-re-
duction of [1,...,1,1,2,1]) can be very time consuming (carry
propagation).

3.5.5.3 Factors of different length

The reduction techniques described in the previous sections can
be generalized readily to factors of different length.

Example. (multiplicand or multiplier possess length 15 or 8 or
vice versa)

```
[0 1 2 3 4 5 6 7 8 8 8 8 8 8 8 7 6 5 4 3 2 1]
                                      <-1 0 3 2>
                  12 8                  6 4 1 1
              <-4 0 12 8>          <-2 0 6 4>
               12 8 1 1              9 6 1 1
            <-4 0 12 8>           <-3 0 9 6>
            10 7 1 1               11 8 1 1
         <-3 0 9 6>             <-4 0 12 8>
          7 5 1 1                12 8 1 1
       <-2 0 6 4>             <-4 0 12 8>
        4 3 1 1                12 8 1 1
    <-1 0 3 2>
 [1 1 1 1 1 1 1 1 1 1 1 1 1 1 1 1 1 1 1 1 1 1]
```

3.5.5.4 On the optimality of the <-1,0,3,2>-element

Consideration of the structure of a multiplication matrix consisting of n lines and k columns shows that the <-1,0,3,2> -element

is optimal, since there is no other element with 8 inputs at most which
requires less modules or a shorter cycle time for the 1-reduction of
the matrix. Perhaps there are modules with a higher number of inputs
with which reduction can be performed more quickly; however, application
of such a module (if at all available) raises problems of cost.

3.6 Pipelining principles

3.6.1 The pipelining concept

The multiplications treated so far do not yet function very efficiently
in the form described, since one step of the instrument at most is
occupied at a given time. In principle, another multiplication could
already be started if the previous multiplication has left the first
step (pipelining principle). This results in increased usage as well
as a considerable increase in speed. However, since cases where
sufficient multiplications are to be carried out simultaneously are
very rare, application of the pipelining concept is not rational in
this extreme form.

It is much more efficient to split an arithmetic operation into several
sub-operations by which the pipelining techniques can be applied.
Figure 3.9 demonstrates how this can be performed with regard to multiplications.

The multiplier is divided into n/h groups of size h. Hence, the
multiplication is split into n/h sub-multiplications with a shorter
multiplier. The h addends of a sub-multiplication are reduced to 2
addends by means of a Wallace-tree (or multiplication according to
Dadda); these are shifted to the right by h bits respectively and
treated further with the addends of the next sub-multiplication. The
period between individual groups of multipliers amounts to r functions
(r = number of functions of the feedback loop). The first group requires

k+r functions for cycling the tree without the final addition (see fig. 3.9), all others require only r additional functions; this yields the following total number of functions z:

$$z = n_h \cdot r + k, \quad (n_h := n/h := \text{number of multiplier groups}).$$

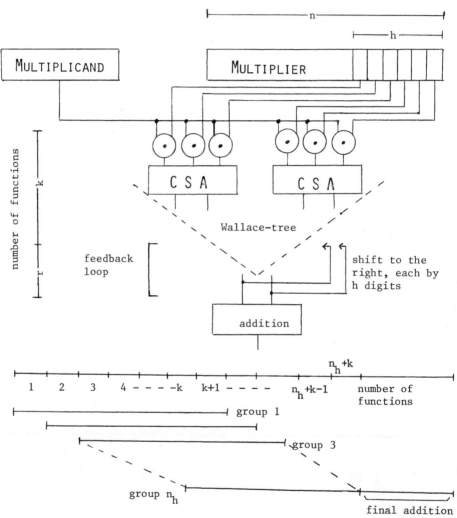

Fig. 3.9 (pipelining multiplication with time schedule of operation for r = 1)

One function per loop (r=1) is sufficient for short feedback loops. In this case we have $z = n_h + k$; see the given time schedule.

Regarding the feedback loop, the h bits on the far right do not need to be carried over since they have no influence on the following groups. Hence, final addition can already be started for these bits. In this way, reduction of different groups, coding of new groups or calculation of new multiples and final addition overlap.

In order to prevent hazards, different steps of this operation must be separated by intermediate storage elements. This has been touched on already by splitting into functions. For temporary storage, special modules ("latches") can be used. A latch blocks the output of a logical element (for example of a CSA adder or a gate) by a simple feedback mechanism. When controlling the latch externally by means of a clock pulse the result is available without further delay. The structure of a latch is described below.

3.6.2 Structure of a latch

3.6.2.1 Latch of a binary data line D

This is the simplest form of latch. It has the following effect:

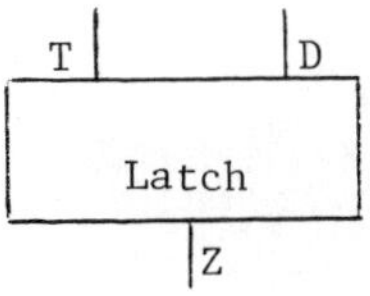

$$Z := \begin{cases} D & \text{where } T = 1 \quad (T = \text{function line}) \\ Z & \text{otherwise.} \end{cases}$$

Fig. 3.10

It is assumed that the value D does not change so long as T = 1 (i.e. in the entire first phase of the function).

The Earle latch ([Ea1], [Ha2]) achieves this by the following formula:

$$Z := \bar{T} \cdot Z \vee D \cdot Z \vee T \cdot D .$$

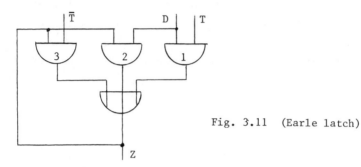

Fig. 3.11 (Earle latch)

Gate 1 places Z on D, as soon as the function line has the value 1 and so long as T = 1, since D does not change during this period by hypothesis. Gate 3 maintains the old value of Z (i.e. D) in the second phase of the function ($\bar{T}$ = 1); during this period, a new value can be calculated for D, adopted in the initial function phase of the next function. Gate 2 avoids hazards between the first and second phase of the function. If T has to be inverted to become $\bar{T}$, there is a slight difference in cycle time between T and $\bar{T}$, and at the beginning of the second phase of the function T as well as $\bar{T}$ may briefly have the value 0.

Omitting gate 2, an error would occur in this case by placing Z at 0, and this value would be maintained.

If $\bar{T} \vee T = 1$ throughout, we could construct the latch very simply, since the relationship $Z := \bar{T} \cdot Z \vee D \cdot Z \vee T \cdot D = \bar{T} \cdot Z \vee T \cdot D$ would apply. With cycle time $\bar{T}$ and T different, it is best to regard $\bar{T}$ and T as two independent variables $\bar{T}_1$ and T_2 :

$$Z := \bar{T}_1 \cdot Z \vee D \cdot Z \vee T_2 \cdot D .$$

3.6.2.2 Construction of latches into logical modules

Latches can be integrated into logical modules without loss of time. This, however, results in more than twice the circuit costs. Furthermore, all gates must be extended by one input (for the function lines T and $\bar{T}$) in the first step of the circuit. Hence, the maximum fan-in of a module is reduced.

We will describe the construction of a latch into a fulladder with 3 inputs a, b, c:

$$s = sum = \bar{a}\bar{b}c \vee \bar{a}b\bar{c} \vee a\bar{b}\bar{c} \vee abc ;$$
$$c = carry = ab \vee ac \vee bc .$$

Sum s* (with built-in latch):

$$s^* = \bar{T} \cdot s^* \vee s \cdot s^* \vee s \cdot T$$
$$= \bar{T} \cdot s^* \vee \bar{a}\bar{b}cs^* \vee \bar{a}b\bar{c}s^* \vee a\bar{b}\bar{c}s^* \vee abcs^* \vee \bar{a}\bar{b}cT \vee \bar{a}b\bar{c}T \vee a\bar{b}\bar{c}T \vee abcT .$$

Carry c* (with built-in latch):

$$c^* = \bar{T} \cdot c^* \vee c \cdot c^* \vee c \cdot T$$
$$= \bar{T} \cdot c^* \vee abc^* \vee acc^* \vee bcc^* \vee abT \vee acT \vee bcT .$$

It is evident that the latch requires a fan-in higher by 1 with regard to AND gates; the cycle time is <u>not increased</u> by the latch (if gates with higher fan-ins are available).

<u>Remark.</u> One function can comprise two latch steps if we interchange the function lines T and $\bar{T}$ in the second step. Hence, the first step takes over in the first phase of the function (T = 1) and the second step in the second phase of the function (T = 0).

3.6.3 Pipelining in conjunction with multiplier coding

Multiplier coding affords a substantial contribution towards increasing the speed of multiplication since coding is performed at the same time as other multiplication steps. We shall describe such a multiplication (see [An2]) for group size 12 of the multiplier and for multiplier coding in groups of 2+1 bits respectively (see 3.3.5). Generalizing to other group sizes and coding rules is obvious.

By multiplier coding, we first obtain 6 multiples of the multiplicand $(s_1,\ldots,s_6)$; these form the inputs of the Wallace-tree. Since both outputs of the tree are fed back, not merely 6 but 8 addends are to be reduced to 2; hence, three levels as shown in table 3.5 are no longer sufficient.

There are various options regarding feedback, for example:

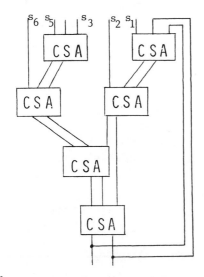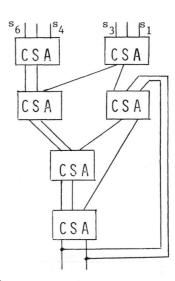

A Total number of levels = 4
   feedback loop of 4 CSA-levels

B Number of levels = 4
   length of loop 3

Fig. 3.12

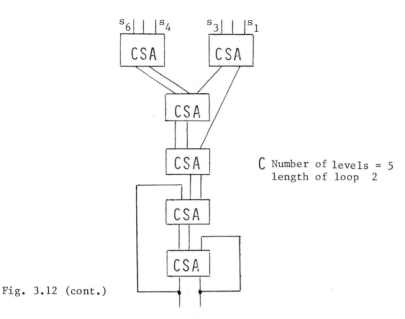

Fig. 3.12 (cont.)

C Number of levels = 5
length of loop 2

Differences in cycle time have to be compensated by delay elements or latches. As described in 3.6.2.2, latches can be incorporated for example in carry save adders. The construction in C, figure 3.12, is especially advantageous because of the short feedback loop, although there is an additional level. In this case, the feedback loop can be traversed in one single function (see note at end of 3.6.2.2).

Figure 3.13 demonstrates the multiplication procedure using the pipelining concept C together with intermediate storage by latches.

Remark. The sequential circuit is split into levels by latches; an attempt is made to split up in such a way so that the number of latches to be incorporated is as small as possible. At the same time, care should be taken that the cycle times necessary for the individual levels do not vary too much (this would act unfavourably as regards frequency

of functions). According to the arrangement chosen in figure
3.13, 4 logical steps at most are processed in each function
(two AND and OR gates each). Below this limit, application
of the pipelining concept is not efficient as regards function
time according to Hallin and Flynn [Ha2].

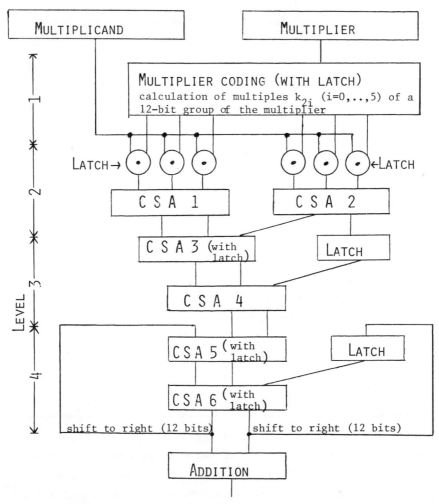

Fig. 3.13 (Pipelining multiplication)

For a multiplier of length 60, coded into 5 groups G1-G5 each of 12 bits, we require $z = 3 + 5 = 8$ functions without final addition of the two remaining addends; this applies irrespective of the length of the multiplicand (see fig. 3.14).

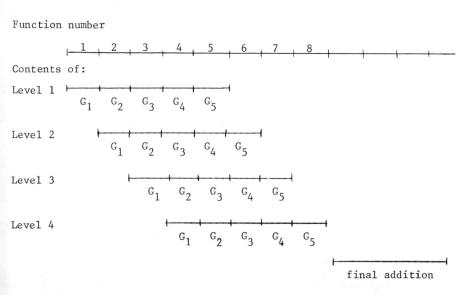

Fig. 3.14 (Pipelining multiplication: time schedule of procedure)

4. Division

4.1 Fundamentals

Division can be understood as the reverse of multiplication. Most division algorithms are based on this interpretation. We will establish the following <u>register configuration</u>:

$(DD, DE) = [DD_{n-1}, ..., DD_0, DE_{n-1}, ..., DE_0]$ dividend register;

$DR = [DR_{n-1}, ..., DR_0]$ divisor register;

$DE = [DE_{n-1}, ..., DE_0]$ quotient register.

The registers DD, DE and DR correspond to multiplication registers MP, MQ and MD (see 3.1).

<u>Notation and scaling</u>

As already discussed in detail regarding multiplication, representation by "amount and sign" is the most suitable (see 3.1); it is best to use the d-complement for negative intermediate results.

Divisor and dividend are regarded as mantissae of floating-point numbers. We will designate both factors by preshifts so that:

$$0 \leq |w(DD)| < |w(DR)|, \text{ and } \frac{1}{d} \leq |w(DR)| < 1.$$

These designations are always possible for $|w(DR)| \neq 0$. An error is indicated, if $w(DR) = 0$, or when the designations result in an overflow of exponents (see 1.6.2).

After carrying out the prepared shift operations, the quotient is in the same area as the dividend; hence, treatment of overflow situations is considerably facilitated.

Recursion formula for division: We invert the corresponding multiplication formula, remembering that we are not working with integral numbers as in 3.2.1, but with mantissae of floating-point numbers $\in (-1:+1)$.

Definition 4.1. $P^{(0)} := 0$

$$P^{(j+1)} := P^{(j)} + d^{j-n+1} \cdot MQ_j \cdot w(MD)$$

$$(j = 0,\ldots,n-1).$$

$$X^{(j)} := \frac{P^{(j)}}{d^{j-n}} \quad (j = 0,\ldots,n).$$

Lemma: *For non-negative multipliers* $(MQ_{n-1} = 0)$, *we have:*

$$X^{(j+1)} = \frac{1}{d} \cdot X^{(j)} + MQ_j \cdot w(MD) ;$$

$$X^{(j)} = d \cdot [X^{(j+1)} - MQ_j \cdot w(MD)]$$

$$(j = n-1, n-2, \ldots, 0).$$

From this formula, we obtain consecutively the digits $MQ_{n-1},\ldots,MQ_0$ of the multiplier (i.e. the quotient bits), based on the "dividend" $X^{(n)} = P^{(n)}$. We thus obtain the following rule of division for non-negative dividends and divisors (writing DR instead of MD):

$$X^{(n)} := w(DD, DE) ;$$

$$q_j := \max \{i \mid 0 \leq i \leq d - 1 \text{ and } X^{(j+1)} - i \cdot w(DR) > 0\} \quad (j=n-1,\ldots,0).$$

$$X^{(j)} := d \cdot [X^{(j+1)} - q_j \cdot w(DR)]$$

$X^{(j)}$ is the shifted, partial remainder (hence also the partial remainder ready for the next recursion step). The remainder of the division thus becomes $X^{(0)}/d$ and $w([q_{n-1},\ldots,q_0])$ is the quotient.

The remainder of the division $R = X^{(0)}/d$ satisfies the conditions:

$|R| < |w(DR)|$ and $\text{sign}(R) = \text{sign}(\text{quotient})$.

In order to uniquely determine the result of the division, we shall require these two characteristics also for non-negative operands.

4.2 Serial division methods for non-negative, binary operands

The recursion formula is simplified with binary base as follows:

$$X^{(n)} := w(DD, DE) \;;$$

$$X^{(j)} := \begin{cases} 2 \cdot X^{(j+1)} & \text{when } X^{(j+1)} - w(DR) < 0, \text{ i.e. } q_j := 0 \\ 2[X^{(j+1)} - w(DR)] & \text{when } X^{(j+1)} - w(DR) \geq 0, \text{ i.e. } q_j := 1. \end{cases}$$

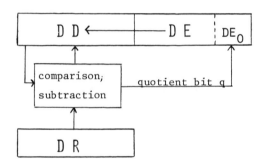

Fig. 4.1

The division algorithm consists mainly of n conditional subtractions of the divisor and of n shifts to the left; note the analogy to multiplication (see fig. 3.1).

4.2.1 Restoring division

Subtraction of w(DR) is always carried out for this variant. If the new partial remainder is negative, it must be cancelled again by addition (restoring). In this case, the quotient bit is 0, otherwise 1.

Microprogram D1 (restoring division)

0 : (DD,DE) := dividend; DR := divisor; Z := n; V := $DD_{n-1} \oplus DR_{n-1}$;

0*: if DD_{n-1} = 1 then (DD,DE) := $\overline{(DD,DE)} + 2^{-2n+1}$;

   if DR_{n-1} = 1 then DR := $\overline{DR} + 2^{-n+1}$;

1 : if $DR_{n-1} = DR_{n-2}$ then [SHL(DD,DE); DE_0:=0; SHL(DR);DR_0:=0;goto 1];

2 : Z := Z-1; DD := DD-DR ;

3 : if DD_{n-1} = 1 then [DD := DD+DR; q := 0] else q := 1 ;

4 : DE_0 := q; if Z > 0 then [SHL(DD,DE); goto 2] else SHL(DE);

4*: if V=1 then [DD := $\overline{DD} + 2^{-n+1}$; DE := $\overline{DE} + 2^{-n+1}$];

5 : END .

Remarks. a. SHL(A) denotes a shift to the left of register A; the position of A on the far left is lost. The character added at the right end depends on the notation chosen; regarding the 2-complement or A+S, a zero is added, if this position is not filled in another manner (see, for example, function 4).

b. The microprogram works for numbers represented in the 2-complement or by A+S. When coding by A+S, the functions marked * are cancelled.

c. Function 1 of the program contains simultaneous shifts of dividend and divisor; these are necessary only if the divisor has not been normalized from the beginning.

d. The division remainder $R = X^{(0)}/2$ is formed by a shift to the right from the last partial remainder $X^{(0)}$; we take this into consideration by relating the last shift to the left of the microprogram to the quotient register DE only and not to the register DD of the partial remainder (function 4). It is evident that the contents of the least significant position DE_0 of the dividend has no influence on the result of the division (rounding). Hence, a dividend register of length 2n-1 (instead of 2n) would be sufficient.

4.2.2 Non-performing division

In this division procedure, subtraction of w(DR) is performed only if the new partial remainder remains non-negative, i.e., if $w(DD) \geq w(DR) > 0$. Here, the time consuming addition/subtraction is replaced by a comparison logic for the contents of registers DD and DR or a subtractor which ceases to function if the sign of the result, usually obtained prior to the result of subtraction, is negative.

<u>Microprogram D2</u> (Non-performing division)

$\vdots$
2 : Z := Z-1; <u>if</u> DD $\geq$ DR <u>then</u>   [q:=1; DD := DD-DR]   <u>else</u> q:=0; goto 4;
$\vdots$

(all other functions as in D1; function 3 cancelled).

4.2.3 Non-restoring division

With regard to the non-restoring division, negative partial remainders are also possible. If a partial remainder is positive, w(DR) is subtracted, otherwise it is added. The quotient bit can be determined from the sign of the new partial remainder. If the partial remainder of the last division cycle is negative, the division remainder must be corrected by adding the divisor.

<u>Non-restoring division, recursion formula:</u>

$$U^{(n)} := w(DD,DE)$$

$$U^{(j)} := \begin{cases} 2[U^{(j+1)} - w(DR)] & \text{when } U^{(j+1)} \geq 0 \\ 2[U^{(j+1)} + w(DR)] & \text{when } U^{(j+1)} < 0 \end{cases} \quad (j=n-1,\ldots,0).$$

$$q_j^U := \begin{cases} 1 & \text{when } U^{(j)} \geq 0 \\ 0 & \text{when } U^{(j)} < 0 \end{cases}$$

We thus obtain the quotient bit q_j^U by inverting the sign of the binary representation of $U^{(j)}$.

Before presenting the respective microprogram, we will demonstrate that the non-restoring method provides the correct quotient:

Lemma 4.1. *If $w(DD,DE) \geq 0$ and $w(DR) > 0$, we have: The quotients $(q_{n-1},\ldots,q_0)$ and $(q_{n-1}^U,\ldots,q_0^U)$, calculated from the recursion formulae for $X^{(i)}$ and $U^{(i)}$ mutually agree. If $U^{(0)}$ is negative, the division remainder must be corrected by adding DR.*

Proof. It is sufficient to demonstrate that:

$$\begin{bmatrix} U^{(k)} = X^{(k)} \geq 0 \\ q_k = q_{k-r-1} = 1 \\ q_{k-1} = \ldots = q_{k-r} = 0 \end{bmatrix} \Rightarrow \begin{bmatrix} U^{(k-r-1)} = X^{(k-r-1)} \geq 0 \\ q_k^U = q_{k-r-1}^U = 1 \\ q_{k-1}^U = \ldots = q_{k-r}^U = 0 \end{bmatrix}$$

Since $q_{k-1} = \ldots = q_{k-r} = 0$, the partial remainders $X^{(k-1)},\ldots X^{(k-r)}$ can be calculated as follows:

$$X^{(k-j)} = 2^j \cdot X^{(k)} \qquad (j = 1,\ldots,r)$$

Furthermore: $X^{(k-j)} < w(DR)$ for $j = 1,\ldots,r-1$.
from $q_{k-r-1} = 1$, we obtain:

$$X^{(k-r-1)} = 2 \cdot [2^r \cdot X^{(k)} - w(DR)] = 2 \cdot [2^r \cdot U^{(k)} - w(DR)] \quad (\geq 0)$$

$$= 2[2^r \cdot U^{(k)} - 2^r w(DR) + 2^{r-1} w(DR) + 2^{r-2} w(DR) + \ldots + 2w(DR) + w(DR)]$$

$$\underbrace{\qquad\qquad\qquad\qquad < 0 \qquad\qquad\qquad\qquad}$$
$$\underbrace{\qquad\qquad\qquad\qquad \geq 0 \qquad\qquad\qquad\qquad}$$

$$= 2[2\{2(\ldots 2(2(U^{(k)}-w(DR))+w(DR))+ \ldots +w(DR))+w(DR)\}+w(DR)]$$

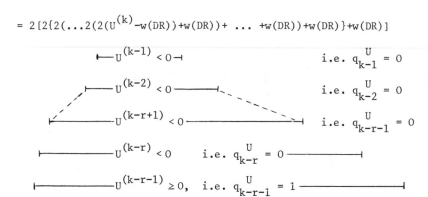

$$\begin{array}{ll} U^{(k-1)} < 0 & \text{i.e. } q^U_{k-1} = 0 \\ U^{(k-2)} < 0 & \text{i.e. } q^U_{k-2} = 0 \\ U^{(k-r+1)} < 0 & \text{i.e. } q^U_{k-r-1} = 0 \\ U^{(k-r)} < 0 & \text{i.e. } q^U_{k-r} = 0 \\ U^{(k-r-1)} \geq 0, & \text{i.e. } q^U_{k-r-1} = 1 \end{array}$$

If $q^U_0 = 0$, the non-restoring division ends with a negative division remainder; there has been one subtraction too many, which must be compensated for by adding the divisor. The quotient bits are not affected by this correction.

<u>Microprogram D3</u>   (Non-restoring division)

0 : (DD,DE) := dividend;   DR := divisor;   Z := n;   V := $DD_{n-1} \oplus DR_{n-1}$;

0*: <u>if</u> DD_{n-1} = 1 <u>then</u> (DD,DE) := $(\overline{DD,DE})$ + 2^{-2n+1} ;

   <u>if</u> DR_{n-1} = 1 <u>then</u> DR := $\overline{DR}$ + 2^{-n+1} ;

1 : <u>if</u> DR_{n-1} = DR_{n-2} <u>then</u> [SHL(DD,DE);SHL(DR);DE_0:=0;DR_0:=0; <u>goto</u> 1]
              <u>else</u> DE_0:= 1 ;

2 : Z := Z-1; <u>if</u> DE_0 = 1 <u>then</u> DD := DD-DR <u>else</u> DD := DD+DR;

3 : $DE_0:=\overline{DD_{n-1}}$; if Z>0 <u>then</u> [SHL(DD,DE); <u>goto</u> 2]
              <u>else</u> [SHL(DE); if DD_{n-1} = 1 <u>then</u> DD :=DD+DR];

4 : <u>if</u> V = 1 <u>then</u> [DD := $\overline{DD}$ + 2^{-n+1}; DE := $\overline{DE}$ + 2^{-n+1}];

5 : END.

<u>Remark.</u> DE_0 always contains the inverted sign of the new partial remainder. At the start of the program, we can fill this position at random, since it does not influence the result of the division.

4.2.4 Shift over zeros and ones

Serial multiplication can be accelerated by shifts over zeros and ones of the multiplier. This applies also to division (shifts over zeros/ones of the partial remainder).

4.2.4.1 Shift over zeros

1. The first quotient bit is always 0, due to the condition $0 \leq w(DD,DE) < w(DR)$. The first subtraction can be replaced by shifting (DD,DE) (over one zero) and entering a zero in DE_0 (this corresponds to the non-performing method).

2. After subtraction (or addition with the non-restoring method), we have:

$$(DD,DE) = \underbrace{00....01}_{k+2}*.....* \qquad (k \geq 0)$$

Hence, the quotient is 1 (since subtraction provides a non-negative partial remainder), and after the respective shift, (DD,DE) starts with k+1 zeros. Since DR = 01*...* , the next k quotient bits at least have the value 0. Hence we can perform another shift (over k positions) and then shift the same amount of zeros in DE.

Considering that the number of shifts is limited by Z+1 (in each case, one quotient bit must be calculated in addition to the number indicated by the present count Z, see microprogram D3: functions 2 and 3) and that the last shift is performed only over register DE, we obtain the following rule:

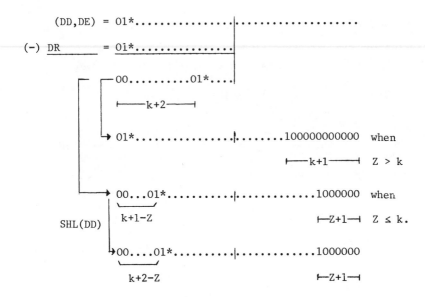

```
        (DD,DE)  = 01*..................|....................
(-)  DR          = 01*................|
                 ┌─ 00..........01*....|
                 │  ├────k+2────┤
                 │
                 └→ 01*................|........100000000000   when
                 │                              ├────k+1────┤   Z > k
                 │
                 └→ 00...01*...........|.............1000000   when
    SHL(DD)      │  ╰──┬──╯                      ├─Z+1─┤       Z ≤ k.
                 │   k+1-Z
                 └→ 00....01*..........|.............1000000
                    ╰──┬──╯                        ├─Z+1─┤
                     k+2-Z
```

The additional shift of DD in the case $Z \leq k$ is compensated for by a shift to the right (only over DD, DD_0 being lost).

4.2.4.2 Shift over ones

Similar to the shift over zeros method, we can also shift over a leading one block of the partial remainder:

```
        (DD,DE)  = 10*..................|....................
(+)  DR          = 01*................|
                 ┌─ 11..........10*....|
                 │  ├────k+2────┤
                 │
                 └→ 10..................|......011111111111   when
                                               ├────k+1────┤   Z > k
```

The simplest way of demonstrating that the quotient bits are
01...1 is by the restoring method (applied for negative dividends
and positive divisors). In this case, subtraction of DR is replaced
by addition, and the quotient bit is 0, if and only if addition could be
performed, i.e., if the new partial remainder is not positive. When
shifting over ones, the first addition can be carried out, since the
new partial remainder has a negative sign; hence, the first quotient bit
is a 0. After performing the first shift, addition is impossible at
least k times since DR = 01*... . Hence, we can shift a further k times
and shift the same amount of ones to DE.

Correction of the remainder. If we obtain the last quotient bit by
shifting over ones, the division remainder is negative; hence it
must be corrected by adding DR. Since the quotient is rounded off
downwards, correction of the quotient is unnecessary.

Remark. With a non-negative dividend/divisor, shifting over ones
is possible only with application of the non-restoring method, since
negative partial remainders do not occur with the two other methods.

4.2.5 Examples of the non-restoring method

1. Application of microprogram D3

(DD,DE) := 010001110010 ;   DR := 0111100 ;

$w(DD,DE) = \dfrac{569}{2^{10}}$; $w(DR) = \dfrac{30}{2^5}$.

DD	DE	
010001	110011 ← DE_0 := 1	
100010		−DR
110011		
100111	10011⌊0	shift
011110		+DR
000101		
001011	0011⌊01	shift
100010		−DR
101101		
011010	011⌊010	shift
011110		+DR
111000		
110000	11⌊0100	shift

DD	DE	
110000	11⌊0100	
011110		+DR
001110		
011101	1⌊01001	shift
100010		−DR
111111		
111111	010010	shift over DE;
011110		+DR (correction,
011101	010010	since partial
⎵ remain- der	⎵ quo- tient	remainder < 0)

$$\frac{569}{2^{10}} : \frac{30}{2^5} = (18+\frac{29}{30}) \times \frac{1}{2^5} \ .$$

2. **Non-restoring with shift over zeros and ones**

$(DD,DE) := 001111011000 \triangleq \frac{492}{2^{10}}$; $DR := 010101 \triangleq \frac{21}{2^5}$

DD	DE	
001111	011001	Z:=5 (DE_0 := 1)
011110	11001⌊0	shift over 0
101011		−DR
001001		Z:=4
010011	1001⌊01	shift
101011		−DR
111110		Z:=3

DD	DE	
111110		
101001	010111	shift over Z+1=4 ones;
110100		SHR (DD only);
010101		+ DR (correction,
001001	010111	since negative partial
⎵ remain- der	⎵ quo- tient	remainder)

$$\frac{492}{2^{10}} : \frac{21}{2^5} = (23 + \frac{9}{21}) \times \frac{1}{2^5} \ .$$

4.3 Negative operands (2-complement notation)

All division methods can be extended to negative dividends/negative divisors. We shall demonstrate this on the example of the non-restoring method:

a. **Partial remainder not identical to zero**

The new partial remainder is calculated according to the formula:

$$DD := \begin{cases} DD - DR & \text{when } DD_{n-1} = DR_{n-1} \\ DD + DR & \text{when } DD_{n-1} \neq DR_{n-1} \end{cases}.$$

For the quotient bit $q = DE_0$, determined in this cycle, we have:

$$DE_0 = \overline{DD_{n-1}} \oplus DR_{n-1}.$$

In this way we obtain exactly the inverted quotient bits (referring to microprogram D3), if either the divisor or the dividend is negative.

Corrections. Because of the 2-complement representation, the result of the division, if negative, must be corrected by adding a 1 to the least significant digit. If division ends in a remainder, the sign of which does not correspond to that of the dividend, correction of the remainder must be performed by adding/subtracting DR. The remainder must be then inverted if it has a sign different from that of the quotient.

b. **Partial remainder identical to zero**

If the partial remainder vanishes, division is "exact"; the remaining quotient bits are determined independently of the sign of the operand as follows:
10.....0   (Z = present count)
├─Z+1─┤

There are no corrections for the division remainder ($\equiv 0$) or of the quotient.

c. Shifts over zeros and ones

If **k** quotient bits $a_1 \ldots a_k$ can be calculated simultaneously (i.e. if the new partial remainder starts with k+1 zeros/ones), the first of these is determined as usual:

$$a_1 = \overline{DD_{n-1}} \oplus DR_{n-1} ;$$

all other a_i (i=2,...,k) have the value $\overline{a_1}$, i.e. $a_1 a_2 \ldots a_k = a_1 \overline{a_1} \ldots \overline{a_1}$.

Microprogram D4 (Non-restoring division for random sign of the operands; without shifts over zeros or ones)

0 : (DD,DE) := dividend; DR := divisor; Z := n;

1 : U := DD_{n-1}; W := DD_{n-1}; X := 1;

  if DR_{n-1} = DR_{n-2} then [SHL(DD,DE); DE_0:=0; SHL(DR); DR_0:=0; goto 1];

1*: if DR = 10...0 then [(DE,DD):= $\overline{(DD,DE)}$ + 0...01; goto 6];

2 : Z := Z-1; if U = DR_{n-1} then DD := DD - DR else DD := DD + DR;

3 : if DD = 0 ∧ DE_{n-1} = ... DE_{n-Z} = 0 then X := 0;

4 : if X = 1

  then [DE_0 := $\overline{DD_{n-1}}$ ⊕ DR_{n-1}; U := DD_{n-1};

    if Z > 0 then (SHL(DD,DE); goto 2)

      else (SHL(DE); if W ≠ DD_{n-1} then

        {if W=DR_{n-1} then DD:=DD+DR else DD:=DD-DR})]

  else [DE_Z := 1; DE_{Z-1} := DE_{Z-2} := ... := DE_0 := 0; Z := 0;

    SHL(DE) over Z+1 digits; goto 6];

5 : if DD_{n-1} ≠ DE_{n-1} then DD := $\overline{DD}$ + 0.....01 ;

  if DE_{n-1} = 1   then DE := $\overline{DE}$ + 0.....01 ;

6 : END .

Remark. a. If the divisor is a negative number with the power two (i.e. DR = 1.11..110000 ≙ 2^{-i} with i > 0) it is normalized by D4 to DR = 1.00..000000 in function 1. In this case, the quotient (≈ - w(DD)) and the remainder can be stated immediately. With regard to the organisation chosen in D4, the signs of remainder and quotient may differ; adjustment of signs is possible without difficulty; we shall not do so because of the special cases.

b. The sign of the partial remainder may be lost by shifting in function 4, hence it is saved in U.

As soon as the partial remainder vanishes (function 3), the variable X is set at 0; in this case, all other quotient bits can be stated immediately (second alternative of function 4).

In the following we shall demonstrate three examples of non-restoring division; in addition to application of microprogram D4, the method of shifting over leading zero/one blocks of the partial remainder is also demonstrated on an example.

Examples

1. Division remainder ≠ 0, without shifts over zeros/ones

Let $w(DD,DE) = \pm \frac{181}{2^8}$; $w(DR) = \pm \frac{14}{2^4}$;

$\frac{w(DD,DE)}{w(DR)} = \pm (12 + \frac{13}{14}) \times \frac{1}{2^4}$.

A	B	C	D
DD:=0101101010	DD:=1010010110	DD:=0101101010	DD:=1010010110
DR:=01110	DR:=01110	DR:=10010	DR:=10010

DD	DE	DD	DE	DD	DE	DD	DE
01011	01010	10100	10110	01011	01010	10100	10110
10010		01110		10010		01110	
11101		00010		11101		00010	
11010	10100	00101	01101	11010	10111	00101	01100
01110		10010		01110		10010	
01000		10111		01000		10111	
10001	01001	01110	11010	10001	01010	01110	11001
10010		01110		10010		01110	
00011		11100		00011		11100	
00110	10011	11001	10100	00110	10100	11001	10011
10010		01110		10010		01110	
11000		00111		11000		00111	
10001	00110	01111	01001	10001	01001	01111	00110
01110		10010		01110		10010	
11111		00001		11111		00001	
11111	01100	00001	10011	11111	10011	00001	01100
01110		10010	+1	01110	+1	10010	corrections
01101	quotient	10011	10100	01101	10100	10011	of the re-
remainder				10010		01100	mainder (R)
				+1		+1	and of the
				10011		01101	quotient

2. Division remainder ≡ 0; with shifts

(DD,DE) := 101101101000   DR := 010101

$$w(DD,DE) = -\frac{588}{2^{10}} = -\frac{21 \times 28}{2^{10}} \qquad w(DR) = \frac{21}{2^5}$$

$$\frac{w(DD,DE)}{w(DR)} = -\frac{28}{2^5} \;\; .$$

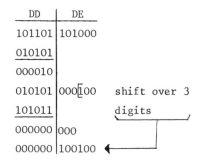

3. Division remainder ≡ 0; without shifts

(DD,DE) = 1100010110   DR = 01001

$w(DD,DE) = -\frac{117}{256}$ $w(DR) = +\frac{9}{16}$

$\frac{w(DD,DE)}{w(DR)} = -\frac{13}{16}$.

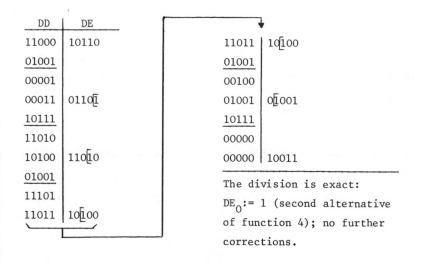

The division is exact:
$DE_0 := 1$ (second alternative
of function 4); no further
corrections.

4.4 Acceleration of division by using suitable multiples of the divisor

4.4.1 Table look up division

In principle, the quotient of two numbers can be calculated from a comprehensive table (see fig. 4.2); division can be reduced to seeking a value in the table (table look up).

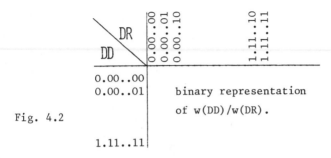

Fig. 4.2

However, since the size of the table increases exponentially with the length of the divisor/dividend, complete application of this method is not realistic for reasons of storing effort and time for looking up on the table. Application of a simplified table containing only the most significant bits of divisor and dividend as inputs is reasonable when calculating several quotient bits in one division cycle.

Cycle procedure in table look up division

1 : Table look up (look up m with $w(DD,DE) \approx m \times w(DR)$ from the table);
2 : $(DD,DE) := (DD,DE) - m \times DR$;
3 : Shift (DD,DE) by α positions to the left and enter the same number of quotient bits. [α is obtained from the length of the leading zero/one block of the new partial remainder or from the count. The quotient bits are determined from the approximation quotient m and the sign of the new partial remainder].
4 : Correction of the quotient bits calculated in earlier cycles; return to 1.

Enlarging the table increases the accuracy of the approximation quotient m, the length of the leading zero/one block of the new partial remainder and the number of the quotient bits determined per cycle. Further corrections are necessary if those bits of the divisor/dividend which were not included in the table influence the quotient bits already calculated.

Serial division methods (4.1 - 4.3) are simply special cases of the table look up division. In this method, only two values are used for m:

$(DD,DE) := (DD,DE) - m \times DR$,

where $m \in \begin{cases} \{+1,0\} & \text{for the restoring/non-performing method} \\ \{+1,-1\} & \text{for the non-restoring division.} \end{cases}$

In the non-restoring method, a cycle consists of one addition/subtraction of the divisor and subsequent normalizing of the partial remainder (by shifting over leading zeros **or** ones). It can be shown that some $\frac{8}{3}$ quotient bits on average are determined per cycle (see Freiman [Fr1]).

4.4.2 Application of special divisor multiples

4.4.2.1 Motivation

Calculation of m x DR or determination of the quotient bits (dependent on m) is extremely tedious (except for special values of m, for example $m = \pm 1, \pm \frac{1}{2},...$). Hence a compromise is sought between complete table-look-up division (all values of m permitted) and serial division methods (only $m = \pm 1$ and $m = 0$ permitted). A divisor multiple m differing from ± 1 is permissible only if

a. calculation of m x DR is easy and quick,
b. the number of quotient bits determined per cycle clearly increases on average (against the mean for serial methods),
c. the quotient bits determined by application of this multiple can be determined by means of a simple rule,
d. no corrections or only simple corrections of the quotient bits calculated in earlier cycles are necessary,
e. it can be readily determined when the multiple m x DR can be used in division.

For reasons of cost it is best to restrict the number of multiples of different value to 3 at most, the value of each of these multiples being larger than, smaller than or equal to 1 respectively.

The number of quotient bits determined in one cycle and the choice of divisor multiple are largely independent of the sign of the operands. Hence it is sufficient to describe methods for positive divisors. Therefore the following tables consider only the value of the divisor; this involves no restriction with representation by amount and sign.

The most significant j bits (except the sign bit) of divisor and dividend are compared with each other. Since the numbers are normalized at the beginning of a cycle and since shifts for dividends of the form $a_1.a_2a_3...a_r...$ or $\overline{a_1}.\overline{a_2}\overline{a_3}...\overline{a_r}...$ are of the same magnitude, the table of the shift sizes of the partial remainder (corresponding to the number of quotient bits determined per cycle) is a $(2^{j-1}) \times (2^{j-1})$-matrix.

MacSorley [Ma1] uses the value j=5. This is still acceptable as regards effort. We shall restrict ourselves in the following to j=4 in order to form smaller and clearer tables. However, this involves no substantial changes.

4.4.2.2 Application of the $\frac{1}{2}$-, 1-, 2-fold of the divisor

4.4.2.2.1 Shift tables

First we shall seek the number of quotient bits which can be determined in one cycle, using the simple divisor and considering 4 positions after the sign respectively. When calculating the size of the shifts we assume that all the following digits of the operands have the value 0. In some rare cases, the shift size can increase or decrease because of a carry from the 5th digit after the sign for operands which do not meet this requirement. Since the table is used only to help choose the multiple of the divisor, this is of little significance.

It can be seen that the size of shifts is constant along the diagonals. The number of shifts is very small when divisor and dividend differ most regarding the amount (bottom left and top right corner of the table).

It is reasonable to apply another divisor multiple in this area. Application of half or double the divisor is appropriate since calculation of these multiples is particularly simple.

Table 4.1 (Sizes of shifts for ± 1 x DR)

DR \ DD	0.1000 / 1.0111	0.1001 / 1.0110	0.1010 / 1.0101	0.1011 / 1.0100	0.1100 / 1.0011	0.1101 / 1.0010	0.1110 / 1.0001	0.1111 / 1.0000
0.1000	≥4	3	2	2	1	1	1	1
0.1001	4	≥4	3	2	2	1	1	1
0.1010	3	4	≥4	3	2	2	1	1
0.1011	2	3	4	≥4	3	2	2	1
0.1100	2	2	3	4	≥4	3	2	2
0.1101	1	2	2	3	4	≥4	3	2
0.1110	1	1	2	2	3	4	≥4	3
0.1111	1	1	1	2	2	3	4	≥4

Examples for table 4.1

a. $+ \begin{cases} DD = 1.0111... \\ DR = 0.1001... \end{cases}$

$\phantom{+ \{DD = }0.0000$

At least four shifts over zeros, if there is no carry from the right, otherwise only 3 shifts:

$$DD = 1.0111101$$
$$DR = 0.1001100$$
$$DD + DR = 0.0001001$$
$$3$$

b. $- \begin{cases} DD = 0.1000... \\ DR = 0.1001... \end{cases}$

$\phantom{- \{DD = }1.1111...$

4 shifts over ones. The number of shifts may increase by a carry from the right:

$$DD = 0.10001101$$
$$DR = 0.10011010$$
$$-DR = 1.01100110$$
$$DD - DR = 0.00000011$$
$$\vdash 6 \dashv$$

In table 4.2 below, the sizes of shifts are indicated using half or double the divisor.

Table 4.2
(Sizes of shifts for $\pm \frac{1}{2}$ x DR (bottom left) or for $\pm$ 2 x DR (top right))

DR \ DD	0.1000 / 1.0111	0.1001 / 1.0110	0.1010 / 1.0101	0.1011 / 1.0100	0.1100 / 1.0011	0.1101 / 1.0010	0.1110 / 1.0001	0.1111 / 1.0000
0.1000	0	1	1	2	2	2	3	4
0.1001	1	0	1	1	1	1	2	2
0.1010	2	1	0	0	1	1	1	1
0.1011	2	1	1	0	0	0	1	1
0.1100	2	2	1	1	0	0	0	0
0.1101	2	2	1	1	1	0	0	0
0.1110	3	2	2	1	1	1	0	0
0.1111	3	2	2	1	1	1	1	0

Examples

a.   DD = 1.0101*
   $\frac{1}{2}$ DR = 0.01101   DR = 0.1101
         1.10111

Size of shift 1;
(increases to 2, if
the next bit of DD
is a 1).

b.   DD = 1.0001
      2DR = 1.0010   DR = 0.1001
            0.0011

Size of shift 2.

A combination of these two tables provides a much higher mean size of shifts than each individually (see table 4.3).

Table 4.3   (Combination of tables 4.1 and 4.2)
<α>: same size of shifts α in table 4.1 and 4.2 .

The areas in which $\pm \frac{1}{2}$ DR or $\pm$ 2DR is used are marked xxx or ooo.

Mean size of shifts:

2.69 bits (restricting all shifts to a maximum of 4 bits).

DR \ DD	0.1000 1.0111	0.1001 1.0110	0.1010 1.0101	0.1011 1.0100	0.1100 1.0011	0.1101 1.0010	0.1110 1.0001	0.1111 1.0000
0.1000	4	3	2	<2>	2	2	3	4
0.1001	4	4	3	2	2	<1>	2	2
0.1010	3	4	4	3	2	2	<1>	<1>
0.1011	<2>	3	4	4	3	2	2	<1>
0.1100	<2>	<2>	3	4	4	3	2	2
0.1101	2	<2>	2	3	4	4	3	2
0.1110	3	2	<2>	2	3	4	4	3
0.1111	3	2	2	2	2	3	4	4

Determination of partial areas for $\pm \frac{1}{2}$ DR or $\pm$2DR is simplified considerably using a multiple deviating from 1 only in the 4 areas at the bottom left and top right. The mean size of shifts is reduced only slightly hereby (see table 4.4):

Table 4.4 (Simplified choice of multiples)

DR \ DD	0.1000 1.0111	0.1001 1.0110	0.1010 1.0101	0.1011 1.0100	0.1100 1.0011	0.1101 1.0010	0.1110 1.0001	0.1111 1.0000
0.1000	4	3	2	2	1 [2]	1 [2]	3	4
0.1001	4	4	3	2	2	1	2	2
0.1010	3	4	4	3	2	2	1	1
0.1011	2	3	4	4	3	2	2	1
0.1100	2	2	3	4	4	3	2	1
0.1101	1 [2]	2	2	3	4	4	3	2
0.1110	3	2	2	2	3	4	4	3
0.1111	3	2	1 [2]	2	2	3	4	4

Shift size α is indicated in the square bracket $\boxed{\alpha}$, as would have been obtained using $\pm \frac{1}{2}$ DR or ± 2 DR.

Mean size of shifts:

   2.62 bits.

The criterion for choosing the divisor multiple is much simpler when using Table 4.4 than for table 4.3 :

$$m = \begin{cases} \pm \frac{1}{2} & \leftrightarrow \quad DR_{n-3}DR_{n-4} = 11 \text{ and } DD_{n-2}DD_{n-3}DD_{n-4} = 011 \text{ or } 100 \\ \pm 2 & \leftrightarrow \quad DR_{n-3}DR_{n-4} = 00 \text{ and } DD_{n-2}DD_{n-3}DD_{n-4} = 000 \text{ or } 111 \\ \pm 1 & \text{otherwise.} \end{cases}$$

4.4.2.2.2 Determination of quotient bits in the ($\frac{1}{2}$, 1, 2)-system

We shall first limit ourselves to non-negative divisors.

A. Application of the simple divisor

Quotient bits do not depend on the sign of the old partial remainder. If the new partial remainder is positive, we enter a 1 and then shift over zeros (entering k-1 zeros, where k denotes the size of shifts). If the new partial remainder is negative, we enter a zero and then shift over k-1 ones.

sign of the new partial remainder	quotient bits
0	1 0 ... 0
1	0 1 ... 1

These rules can be proved by applying one of the division methods dealt with in 4.2.1 - 4.2.3 (restoring, non-performing or non-restoring division).

B. Application of the half divisor

B.1 Old partial remainder positive

The divisor is subtracted. Consideration of table 4.4 demonstrates that in all cases in which half the divisor was applied, the partial remainder keeps its sign (**) and that subtraction of the full divisor would invert the sign (*). After subtracting $\frac{1}{2}$ DR we can still shift over leading zeros of the new partial remainder (***):

Quotient bits: 0   1   0 ... 0 .
               (*) (**) (***)

B.2 Old partial remainder negative

In a similar manner to B.1 it is shown that the quotient bits are 1 0 1 .... 1 (the new partial remainder is negative).

C. Application of the double divisor

C.1 Old partial remainder positive

The double divisor is used only for very large partial remainders and very small divisors. When subtracting the simple divisor, the remainder would stay positive. Hence, the first quotient bit is a 1. The following quotient bits are also 1; we can see this immediately if we determine the quotient bits by the non-restoring method for the 4 cases, in which 2 x DR is used. Since the new partial remainder is negative when subtracting the double divisor, the quotient bits correspond to the sign of the new partial remainder.

C.2 Old partial remainder negative

Similar considerations show that the quotient bits are 0 ..... 0 . They correspond to the sign of the new partial remainder.

These rules can be readily applied for negative divisors:

Table 4.5 (Quotient bits, applying $(\pm \frac{1}{2}, \pm 1, \pm 2) \cdot DR$)

m	quotient bits
$\pm \frac{1}{2}$	$\bar{a}aaa\ldots\ldots\ldots a$
± 1	$\bar{a}aaa\ldots\ldots\ldots a$
± 2	$\bar{a}aaa\ldots\ldots\ldots a$
	⊢size of shifts⊣

where $a := (u \oplus DR_{n-1}) \cdot X$;

$u :=$ sign of the new partial remainder;

$$X := \begin{cases} 0 \text{ where partial remainder} \equiv 0 \\ 1 \text{ where partial remainder} \not\equiv 0 \end{cases}$$

If division is exact (X = 0 after the last division cycle), we can dispense with corrections; otherwise, quotient and remainder must be corrected as in the non-restoring method (see 4.3).

Examples:

1. (DD,DE) := 1100010110   DR := 01001

    $w(DD,DE) = -\frac{117}{2^8}$ $w(DR) = +\frac{9}{2^4}$

    $\frac{w(DD,DE)}{w(DR)} = -\frac{13}{2^4}$

DD	DE	
11000	10110	
10001	0110*1*	shift over a 1
10010		+ 2 · DR
00011		
01101	10*1*00	shift over two zeros
10111		− 1 · DR
00100		
01001	0*1*001	shift over one zero
10111		− 1 · DR
00000		
00000	10011	shift over DE only

2. (DD,DE) := 00100010010100   DR := 1000011

$$w(DD,DE) = \frac{1098}{2^{12}} \qquad w(DR) = -\frac{61}{2^6}$$

$$\frac{w(DD,DE)}{w(DR)} = -\frac{18}{2^6}$$

DD	DE	
0010001	0010100	
0100010	010100⌊1	shift over 1 zero
1100001	1	$+ \frac{1}{2}$ x DR
0000011	1	
0111101	00⌊11011	shift over 4 zeros
1000011		+ 1 x DR
0000000		
0000000	1101110	

<u>Remark.</u> In order to avoid rounding errors, the first position of DE must be included in the arithmetic when applying the half divisor (if $DR_0 = 1$); see example 2.

4.4.2.3 <u>Application of the $\left(\frac{3}{4}-,\ 1-,\ \frac{3}{2}\right)$- divisor</u>

The disadvantage of the method discussed in 4.4.2.2 is that the divisor multiples applied which deviate from ± 1 do not achieve their optimum effect (maximum size of shifts) until they are in the "corners" of the table. Multiples which would achieve their optimum value in the middle between the main diagonal and the corners would be more suitable. If we restrict ourselves to 3 divisor multiples which differ regarding their amount and if we assume that all values in the table have the same probability (usually this condition is not exactly met, but there are only slight deviations), we obtain a maximum size of shifts, using the $(\frac{77}{128},\ 1,\ \frac{53}{32})$-divisor (see [Fr1]).

Due to the requirement of 4.4.2.1, these values are, of course, excluded. The $(\frac{3}{4}, 1, \frac{5}{4})$-system is nearly as good and comes closer to the requirements. Since the 3- and 5-multiple of the divisor must be calculated separately in two different additions for this system, we shall not discuss this here.

Above all, the $(\frac{3}{4}, 1, \frac{3}{2})$-system and the $(\frac{5}{8}, 1, \frac{5}{4})$-system are valid with regard to application. We shall restrict ourselves to the former of the two since it offers slightly easier determination of the quotient bits (furthermore, the same considerations would apply generally for the second system). The $(\frac{3}{4}, 1, \frac{3}{2})$-system was used when constructing the STRETCH-machine (see [Bu1]).

4.4.2.3.1 Shift tables

Table 4.6

Sizes of shifts for $\pm \frac{3}{4}$ DR (bottom triangle) or $\pm \frac{3}{2}$ DR (top triangle). $\pm 1 \times$ DR is used on the main diagonal.

DR \ DD	0.1000 1.0111	0.1001 1.0110	0.1010 1.0101	0.1011 1.0100	0.1100 1.0011	0.1101 1.0010	0.1110 1.0001	0.1111 1.0000
0.1000	*	2	3	4	4	3	2	2
0.1001	2	*	2	3	3	4	3	2
0.1010	3	2	*	2	2	3	4	4
0.1011	4	3	2	*	2	2	3	4
0.1100	4	4	3	2	*	1	2	2
0.1101	4	4	3	2	2	*	1	2
0.1110	3	4	4	3	2	2	*	1
0.1111	2	3	4	4	3	2	2	*

The table 4.7 below indicates the sizes of shifts, using the $\frac{3}{4}$- or $\frac{3}{2}$- divisor.

Having in mind determination of cases as simple as possible using $\pm\frac{3}{4}$ DR or $\pm\frac{3}{2}$ DR, we shall determine a table with a mean size of shifts only slightly reduced (table 4.8); this construction corresponds to developing table 4.3 from tables 4.1 and 4.2 .

Table 4.7 (Maximum size of shifts in the $(\frac{3}{4}, 1, \frac{3}{2})$-system

<α>: the same size of shifts α for two different divisor multiples.

Mean size of shifts

(restricting all shifts to a maximum of 4 bits): 3.33 bits .

DR \ DD	0.1000 1.0111	0.1001 1.0110	0.1010 1.0101	0.1011 1.0100	0.1100 1.0011	0.1101 1.0010	0.1110 1.0001	0.1111 1.0000
0.1000	4	3	3	4	4	3	2	2
0.1001	4	4	3	3	3	4	3	2
0.1010	<3>	4	4	3	<2>	3	4	4
0.1011	4	<3>	4	4	3	<2>	3	4
0.1100	4	4	<3>	4	4	3	<2>	<2>
0.1101	4	4	3	3	4	4	3	<2>
0.1110	3	4	4	3	3	4	4	3
0.1111	2	3	4	4	3	3	4	4

The areas where $\pm\frac{3}{4}$ x DR or $\pm\frac{3}{2}$ x DR is used are marked xxx or ooo .

Table 4.8 (see following page)

☒: size of shift α would be possible by choosing a different multiple.

Mean size of shifts

   3.28 bits.

DR \ DD	0.1000 / 1.0111	0.1001 / 1.0110	0.1010 / 1.0101	0.1011 / 1.0100	0.1100 / 1.0011	0.1101 / 1.0010	0.1110 / 1.0001	0.1111 / 1.0000
0.1000	4	3	3	4	4	3	2	2
0.1001	4	4	3	2 ③	3	4	3	2
0.1010	3	4	4	3	2	3	4	4
0.1011	4	3	4	4	3	2	3	4
0.1100	4	4	3	4	4	3	2	2
0.1101	4	4	3	2 ③	4	4	3	2
0.1110	3	4	4	3	3	4	4	3
0.1111	2	3	4	4	3	2 ③	4	4

4.4.2.3.2 Determination of the quotient bits

More case differentiations must be carried out in the $(\frac{3}{4}, 1, \frac{3}{2})$-system than in the $(\frac{1}{2}, 1, 2)$-system. The quotient bits are obtained from table 4.9. The indications obtained can be readily checked; see investigations below:

A. **Addition of $-\frac{3}{4}$ DR (w(DR) > 0):**

The old partial remainder was positive (otherwise $+\frac{3}{4}$ DR would be added). The easiest way of proving the rules is by applying the non-performing method:

Addition of $-1 \times$ DR would invert the sign of the partial remainder in all cases where $-\frac{3}{4}$ DR is used (see table 4.8); hence, addition is not performed and the first quotient bit is zero.

Addition of $-\frac{3}{4}$ DR can be divided into two parts:

1. Addition of $-\frac{1}{2}$ DR : the sign of the partial remainder is maintained; the 2nd quotient bit has the value 1.

2. Addition of $-\frac{1}{4}$ DR : if the partial remainder continues positive the third quotient bit is 1 and can still be shifted over zeros (line ③ of table 4.9) ;

if it becomes negative, the third bit is 0 which is followed by an eventual shift over ones afterwards (line ④).

Table 4.9 (Quotient bits in the $(\frac{3}{4}, 1, \frac{3}{2})$-system)

application of	$(u \oplus DR_{n-1}) \times X$	quotient bits	
① $\pm$ DR	0	1000...0	
②	1	0111...1	
③ $-\frac{3}{4}$ DR	0	0110...0	
④	1	0101...1	(*)
⑤ $+\frac{3}{4}$ DR	0	1010...0	
⑥	1	1001...1	
⑦ $-\frac{3}{2}$ DR	0	1100...0	
⑧	1	1011...1	
⑨ $+\frac{3}{2}$ DR	0	0100...0	
⑩	1	0011...1	

[u and X are defined as in table 4.5].

(*) : with the size of shifts being 2, the first of the quotient bits determined in the next cycle must be inverted.

B. Addition of $-\frac{3}{2}$ DR (w(DR) > 0):

We again divide the operation into two parts:

1. Addition of $-$ 1 DR : the sign of the partial remainder continues unchanged in all cases, hence the first quotient bit is 1.

2. Addition of $-\frac{1}{2}$ DR : the following quotient bits are 10....0 or 01....1, depending on the new partial remainder being positive (line ⑦) or negative (line ⑧).

The arguments follow accordingly in the remaining cases (addition of $+\frac{3}{4}$ DR or of $+\frac{3}{2}$ DR).

C. <u>Negative divisors</u>

We shall examine what changes occur as compared with the results for the positive divisor of the same value:

1. Instead of adding m x DR, -m x DR is added.
2. All quotient bits are to be inverted.

Hence the quotient bits of lines ③ and ⑥, ④ and ⑤, ⑦ and ⑩, ⑧ and ⑨ must be interchanged in accordance with the rules for positive divisors; at the same time, however, the value of $u \oplus DR_{n-1}$ is inverted. Both interchanges mutually compensate; hence, table 4.9 also applies for negative divisors.

<u>Remark</u>. Attention should be paid so that the quotient bits reflect the total divisor multiple (independently of the actual size of shifts). Hence, when using $\pm\frac{3}{2}$ DR, at least 2 quotient bits per cycle shall be determined, and when using $\pm\frac{3}{4}$ DR, at least 3 shall be determined. Consideration of all cases shows that the minimum size of shifts cannot fall to 1 even by carries from the right (see argument in 4.4.2.2.1). Hence there are no problems for $\pm\frac{3}{2}$ DR . Regarding $\pm\frac{3}{4}$ DR and size of shift 2, the number of quotient bits determined per cycle is too small by 1. Addition of $\pm\frac{1}{4}$ DR cannot be considered until the next cycle (see correction rules in table 4.9 and the following examples).

D. <u>Division remainder $\equiv$ 0</u>

If division is exact, all other quotient bits can be stated immediately. It can be readily considered that the first alternative be chosen, independently of the sign of the divisor or dividend in table 4.9 (lines ①, ③, ⑤, ⑦, ⑨); possible corrections of quotients are included.

Application of the division method (using the $(\frac{3}{4}, 1, \frac{3}{2})$-fold divisor) is demonstrated on three examples:

Examples

1. (DD,DE) := 01101010111000   DR := 1000100

    $w(DD,DE) = \frac{3420}{2^{12}}$ $w(DR) = -\frac{60}{2^6}$

    $\frac{w(DD,DE)}{w(DR)} = -\frac{57}{2^6}$

    $+\frac{3}{4}$ DR = 101001100 $-\frac{3}{4}$ DR = 010110100

DD	DE	
0110101	0111000	
1010011	00	$+\frac{3}{4}$ DR
0001000		
0100001	110001̄0 (0)	(line ⑥)
1010011	00	$+\frac{3}{4}$ DR
1110100		
1010011	0001̄01̄0 (1)	(line ⑤)
0101101	00	$-\frac{3}{4}$ DR
0000000		
0000000	101̄001̄1	(line ③)
	1000111 =	
	quotient	

2. (DD,DE) := 1101110 1101100   DR := 1000011

    $w(DD,DE) = -\frac{1098}{2^{12}}$ $w(DR) = -\frac{61}{2^6}$

    $\frac{w(DD,DE)}{w(DR)} = \frac{18}{2^6}$

$+\frac{3}{4}$ DR = 101001001 $-\frac{3}{4}$ DR = 010110111

DD	DE
1101110	1101100
1011101	1011000
0101101	11
0001011	01
0101101	1100001 (0)
1010010	01
0000000	00
0000000	0011010
	0010010

shift over 1
$-\frac{3}{4}$ DR

(line ④)
$+\frac{3}{4}$ DR

(line ⑤)
=

In order to avoid rounding errors, the first two bits of DE must be included in the arithmetic (if $DR_0 = 1$).

3. (DD,DE) := 1100010110   DR := 01001

$w(DD,DE) = -\frac{117}{2^8}$ $w(DR) = +\frac{9}{2^4}$

$\frac{w(DD,DE)}{w(DR)} = -\frac{13}{2^4}$ $+\frac{3}{2}$ DR = 011011

DD	DE
11000	10110
10001	01101
01101	1
11110	1
10111	01001
01001	
00000	
00000	10011

shift over one 1
$+\frac{3}{2}$ DR

(line ⑩)
+ DR

(line ①)

4.5 Iterative division

4.5.1 Motivation

A division technique differing completely from the methods discussed so far consists of calculating the quotient by repetition of simpler operations (addition/multiplication); this is called "iterative division". The following factors support this method:

1. Serial division methods are more complicated than the respective multiplication methods, particularly as regards negative operands.

2. Since division is generally less frequent than multiplication, any time loss due to iterative methods would have hardly any effect.

3. Application of the fastest multiplication as well as the fastest division methods is often prohibitive for reasons of cost. However, a fast multiplication algorithm is more favourable as regards costs, if it can be used also for other operations.

4. The principles of the iterative division can be applied to other operations also (for example extraction of the root, see 4.5.5).

5. The speed of iterative methods is competitive with the speed of other methods. Among other things, we have the following accelerating options:

a. Table look up for finding a favourable starting value;
b. shorter (i.e. approximate) calculation of intermediate results during the first steps of iteration;
c. parallel execution of various operations (pipelining).

4.5.2 Notation

Let dividend and divisor be <u>non-negative</u> and normalized, n-digit mantissae of binary floating point numbers of the form $a_{n-1} \cdot a_{n-2} \ldots a_0$, where $w(a_{n-1} \cdot a_{n-2} \ldots a_0)$ is defined by:

$$w(a_{n-1} \cdot a_{n-2} \ldots a_0) := \sum_{i=0}^{n-1} a_{n-1-i} \cdot 2^{-i} \in [0 : 2 - \frac{1}{2^{n-1}}] \ .$$

Result. *a. Calculation of representation of 1±α from representation of 1∓α (0 ≤ α ≤ 1) (or representation of 2-α from that of α) is performed by inverting all bits and adding a 1 to the least significant digit.*

b. If α < 1, then 1-α can be calculated by inverting all bits except the first and adding a 1 to the least significant digit.

Example: α ≙ 0.0110100 ⇒ 2-α ≙ 1.1001100 ; 1-α ≙ 0.1001100 .

4.5.3 Rules of iteration, convergence of iteration methods

Definition 4.2. *Assume a case with real values φ.*

a. Calculation of the sequence $x_0, x_1, x_2, \ldots$ by means of

$$x_{i+1} := \varphi(x_i) \qquad (i \geq 0)$$

is called iteration method *with starting value x_0. φ means* rule of iteration.

b. If φ is sufficiently capable of differentiation, the iteration method (choosing a suitable starting value) converges with order m *towards the* limit value $\bar{x}$ *(in symbols: $x_i \xrightarrow[i \to \infty]{} \bar{x}$), when:*

$$\varphi(\bar{x}) = \bar{x}; \quad \varphi'(\bar{x}) = \varphi''(\bar{x}) = \ldots = \varphi^{(m-1)}(\bar{x}) = 0; \quad \varphi^{(m)}(\bar{x}) \neq 0 .$$

c. $\varepsilon_j := x_j - \bar{x}$ signifies error *in the j-th step of iteration.*

There is the following relationship between iteration error and order:

Lemma 4.2. $x_{i+1} := \varphi(x_i)$ *converges with order m exactly if we have:*

$$\varepsilon_{j+1} = O(\varepsilon_j^m) \qquad (j \geq 0).$$

Proof. By the Taylor development of φ around the limit value $\bar{x}$, we obtain:

$$\varepsilon_{j+1} = x_{j+1} - \bar{x} = \varphi(x_j) - \bar{x} = \varphi(\bar{x}+\varepsilon_j) - \bar{x}$$

$$= \sum_{i=0}^{\infty} \frac{\varphi^{(i)}(\bar{x})}{i!} \cdot \varepsilon_j^i - \bar{x} = \sum_{i=1}^{\infty} \frac{\varphi^{(i)}(\bar{x})}{i!} \cdot \varepsilon_j^i .$$

This yields:

$$\varphi^{(i)}(\bar{x}) = 0 \quad (i=0,\ldots,m-1) \quad \Leftrightarrow \quad \varepsilon_{j+1} = O(\varepsilon_j^m) .$$

Examples:

1. Newton method:

$$\varphi(x) := x - \frac{f(x)}{f'(x)} .$$

If $f'(\bar{x}) \neq 0$ and if f is continuously capable of differentiation twice, there is a square convergence of the rule of iteration defined by φ (i.e. to the order 2) towards one zero digit $\bar{x}$ of $f(x)$. For the iteration error, we have:

$$\varepsilon_{i+1} = \frac{f''(\xi)}{2f'(x_i)} \cdot \varepsilon_i^2 , \text{ where } \xi \in [\bar{x}, x_i] .$$

The conditions for fast convergence are (this applies to other methods in the appropriate manner):

a. Starting value x_0 sufficiently close to $\bar{x}$;
b. f'' limited and relatively small in $[\bar{x}, x_0]$;
c. $f'(x_i)$ not so small compared with $f''(\bar{x})$.

2. Generalized Newton method:

$$\varphi(x) := x - \frac{f(x)}{f'(x)} - \frac{1}{2} \cdot \left(\frac{f(x)}{f'(x)}\right)^2 \cdot \frac{f''(x)}{f'(x)} ;$$

3. **Halley iteration:**

$$\varphi(x) := x - \frac{\frac{f(x)}{f'(x)}}{1 - \frac{f(x)}{f'(x)} \cdot \frac{f(x)}{f''(x)}} ;$$

4. **Bailey iteration:**

$$\varphi(x) := x - \frac{f(x)}{f'(x) - \frac{f(x) \cdot f''(x)}{2 \cdot f'(x)}} .$$

Methods 2. - 4. are of the order 3.

4.5.4 Iterative division methods

Iterative methods for calculating quotients ascribe division to a sequence of simpler operations (addition, multiplication, shifts etc.). The rule of iteration must be "simple" (regarding the number of operations to be carried out per step of iteration) and free of division. It is sufficient to calculate the inverted value $\frac{1}{b}$ of the divisor instead of the quotient $\frac{a}{b}$. We then obtain the quotient by final multiplication with a.

In the following it is assumed that a and b are normalized mantissae of floating-point numbers for which we have: $\frac{1}{2} \leq a, b < 1$.

4.5.4.1 Division according to Newton

For the Newton method (see 4.5.3), we define f(x) as follows:

$$f(x) := \frac{1}{x} - b,$$

i.e. $\varphi(x) := 2x - b \cdot x^2 = (2 - b \cdot x) \cdot x$.

The division free iteration method resulting from this provides a square convergence towards the zero digit $\frac{1}{b}$ of $f(x)$.

> **Newton method (2nd order):**
>
> $$x_0 := 1 \, ; \quad x_{i+1} := (2 - b \cdot x_i) \cdot x_i \, .$$

Two multiplications and calculation of $2-\alpha$ from α are necessary per iteration step.

Iteration errors of the Newton method:

$$\varepsilon_{i+1} = x_{i+1} - \bar{x} = (2-bx_i) \cdot x_i - \frac{1}{b} = -b \cdot \left(x_i - \frac{1}{b}\right)^2$$

$$= -b \cdot \varepsilon_i^2 \le 0 \, .$$

Without the inevitable rounding errors, the method would converge <u>from below</u> towards $\frac{1}{b}$.

In order to accelerate division it is advisable to start with a table look up value for $\frac{1}{b}$ (for example by inspecting the first k bits of the b-representation, see 4.5.4.3).

4.5.4.2 Generalized Newton iteration and division according to Ferrari [Fe1]

The Newton method can be generalized to higher orders. We can increase the order by 1 by means of an additional multiplication per iteration. The rule of iteration then reads:

> **Generalized Newton method (order m+1):**
>
> $x_0 :=$ table look up value for $\frac{1}{b}$ (or $x_0 := 1$) ;
>
> $r_{i+1} := 1 - b \cdot x_i; \quad x_{i+1} := (1 + r_{i+1} + r_{i+1}^2 + \ldots + r_{i+1}^m) \cdot x_i$
>
> $\qquad \qquad \qquad = (1 + r_{i+1} \cdot (1 + r_{i+1} \cdot (1 + \ldots \cdot (1 + r_{i+1}) \ldots) \cdot x_i \, .$
>
> $\underbrace{\qquad}_{\text{1 mult.}} \quad \underbrace{\qquad\qquad\qquad\qquad}_{\text{m multiplications}}$

In the special case $m=1$, we obtain the Newton method from 4.5.4.1.

Lemma 4.3. *For the generalized Newton method, we have:*

$\left. \begin{array}{l} r_i \text{ converges towards zero} \\ x_i \text{ converges towards } \frac{1}{b} \end{array} \right\}$ *convergence order* $m+1$.

<u>Proof.</u> We will first demonstrate that $r_{i+1} = r_i^{m+1}$ (i.e. $r_i \to 0$ with order $m+1$); since:

$$1 - r_i = b \cdot x_{i-1} = b \cdot \frac{x_i}{1 + r_i + \ldots + r_i^m}$$

$$\Rightarrow (1 - r_i)(1 + r_i + \ldots + r_i^m) = b \cdot x_i = 1 - r_{i+1}$$

i.e. $1 - r_i^{m+1} = 1 - r_{i+1}$,

hence

$$r_i^{m+1} = r_{i+1}.$$

Furthermore, we have:

$$r_{i+1} = 1 - b \cdot x_i = b \cdot (\frac{1}{b} - x_i) = -b(x_i - \bar{x}) = -b \cdot \varepsilon_i \; ;$$

i.e. $\varepsilon_{i+1} = -\frac{r_{i+2}}{b} = -\frac{r_{i+1}^{m+1}}{b} = (-b)^m \cdot \varepsilon_i^{m+1}$;

this concludes the proof.

A generalized Newton method of order $m+1$ is called optimal if it provides the highest accuracy with given effort (determined by the total number of multiplications performed). It can be readily shown that the Newton method for order 3 (i.e. $m=2$) is optimal with regard to this interpretation.

> **Ferrari method (generalized Newton method for order 3):**
> $x_0 :=$ table look up value for $\frac{1}{b}$;
> $r_{i+1} := 1 - b \cdot x_i$; $x_{i+1} := (1 + r_{i+1} \cdot (1 + r_{i+1})) \cdot x_i$.

4.5.4.3 Anderson-Earle-Goldschmidt-Powers method [An2]

The (generalized) Newton methods for order m+1 have the disadvantage that all m+1 multiplications are sequential and cannot therefore be performed simultaneously.

The following square convergence method – developed for the IBM 360/91 – has the advantage that both multiplications of one iteration step can be carried out in parallel. Furthermore, it converges towards the quotient $\frac{a}{b}$ and not towards the inverted value $\frac{1}{b}$ of the divisor.

> **Anderson-Earle-Goldschmidt-Powers method:**
> $d_0 := b$; $\qquad\qquad x_0 := a$;
> $d_{i+1} := d_i \cdot (2-d_i)$; $\qquad x_{i+1} := x_i \cdot (2-d_i)$.

It is evident that this is a variant of the Newton method; hence, the following characteristics emerge:

1. $d_i \xrightarrow[i \to \infty]{} 1$; the convergence is square (for proof, see Newton method (4.5.4.1) and write b=1).

2. Since $x_j/d_j = x_0/d_0$ (this can be readily shown by induction), there is a square convergence of the sequence of x_i towards $x_0/d_0 = a/b$.

3. Both sequences converge <u>from below</u> towards their limit values (if there is no rounding error).

4. 2 multiplications are necessary per step; these can be performed <u>in parallel.</u>

5. For factors of word length n, $\lceil \log_2 n \rceil$ iterations are necessary at most, i.e. within the achievable accuracy, we have:

$$d_E = 1; \quad x_E = \frac{a}{b}, \quad \text{where} \quad E = \lceil \log_2 n \rceil \;.$$

Some iterations can be saved by a different choice of starting values (table look up).

6. Since it is a self correcting method (it also converges for imprecise iteration values d_{i+1} or x_{i+1} towards the same limit values), it is sufficient to calculate d_{i+1} and x_{i+1} approximately. For example, we can use the same shortened multiplier $(2-d_i)_T$ instead of $(2-d_i)$ for $d_{i+1} = d_i \cdot (2-d_i)$ as well as for $x_{i+1} = x_i \cdot (2-d_i)$. Hence, the speed of multiplication which is decisively influenced by the length of the multiplier, can be increased further and the time necessary for division can be reduced considerably.

7. Only in the last iteration (calculation of $x_E = x_{E-1} \cdot (2-d_{E-1})$) need we use the full length of the multiplier. However, we do not have to determine d_E in this step, since this auxiliary value is no longer necessary and since it has the value 1 within the achievable accuracy. Hence we can arrange multiplication in such a way so that two shortened multiplications (for the calculation of d_i or x_i) are performed in parallel during the first E-1 iterations. It is used for multiplication of normal length only for the last iteration.

In the following we shall discuss some of the previously mentioned possibilities for improving the Anderson method:

I. Speed of convergence and direction of convergence

It can readily be shown by induction that:

$$d_i = 1 - x^{2^i}, \quad \text{where} \quad x = 1 - b \;.$$

(Proof: $d_i = 1 - x^{2^i} \Rightarrow d_{i+1} = d_i \cdot (2-d_i) = (1-x^{2^i}) \cdot (1+x^{2^i}) = 1-x^{2^{i+1}}$).

Since b is assumed normalized ($\frac{1}{2} \leq b < 1$), we have $0 < x \leq \frac{1}{2}$; i.e. $d_h \in [1-(\frac{1}{2})^{2^h} : 1]$.

The binary representation of d_h has the following form

$$d_h \triangleq \underbrace{\varepsilon.\bar{\varepsilon}.....\bar{\varepsilon}}_{2^h}*.....* \;, \quad \varepsilon \in \{0,1\}$$

This yields:

$$d_{\lceil \log_2 n \rceil} \triangleq \underbrace{\varepsilon.\bar{\varepsilon}..........\bar{\varepsilon}}_{\substack{\geq n \\ \geq n+1}}$$

With regard to registers of length n, at least one digit must be rounded (upwards, when $\varepsilon = 0$, downwards, when $\varepsilon = 1$); Hence:

$$d_{\lceil \log_2 n \rceil})_{\text{rounded}} \triangleq \underbrace{1.0.....0}_{n}$$

Calculating exactly (i.e. without rounding errors), we obtain $\varepsilon = 0$, i.e. convergence from below (because of $d_i = 1-x^{2^i} < 1$; see 4.5.4.1). The direction of convergence can change due to rounding errors, by approximate calculation of the iterated values (see III.), or by using a table look up value at the start of iteration; in this case, $\varepsilon = 0$ as well as $\varepsilon = 1$ are possible.

II. <u>Acceleration by table look up</u>

Convergence is increasingly accelerated towards the end of iteration; the (h+1)-th iteration step alone increases the accuracy (number of correct binary digits in the result) by the same amount as the combined initial h iterations:

$$d_h \triangleq \underbrace{\varepsilon.\bar{\varepsilon}.....\bar{\varepsilon}}_{2^h}*.....* \;\; ; \;\; d_{h+1} \triangleq \underbrace{\varepsilon.\bar{\varepsilon}..........\bar{\varepsilon}}_{2 \cdot 2^h}*.....* \;.$$

We therefore attempt to combine the initial, still relatively "ineffective", iterations:

Iteration method with table look up:

$y := f(b) \approx \frac{1}{b}$;

$d_1 := b \cdot y$; $x_1 := a \cdot y$;

$d_{i+1} := d_i \cdot (2-d_i)$; $x_{i+1} = x_i \cdot (2-d_i)$.

In this case, y is an approximate value for $\frac{1}{b}$, obtained by inspection of the most significant bits of the binary representation of b. Considering k bits, we can achieve that $|1-d_1| < 2^{-k}$:

$$b \triangleq 0.1*\ldots*\,*\ldots* \atop \vdash\!\!-k\!\!-\!\!\dashv$$

$$y \triangleq 1.*\ldots\ldots\ldots*$$

$$d_1 = b \cdot y \triangleq \begin{cases} 0.1\ldots\ldots 1*\ldots\ldots* \\ 1.0\ldots\ldots 0*\ldots\ldots* \end{cases}.$$
$$\vdash\!\!-k\!\!-\!\!\dashv$$

Instead of $\lceil \log_2 n \rceil$, only $\lceil \log_2 \frac{n}{k} \rceil + 1$ iterations are necessary.

Example: n=56, k=7 (Table with 128 inputs).

Without table look up: $\lceil \log_2 56 \rceil$ = 6 iterations.

With table look up: $\lceil \log_2 8 \rceil + 1$ = 4 iterations (after the individual iteration steps, the deviation from the limit value is less than $2^{-7}, 2^{-14}, 2^{-28}$ or 2^{-56}).

III. **Shortening of the multiplier (approximate calculation of iterations)**

For binary representation of d_i, we have:

$$d_i \triangleq \epsilon.\overline{\epsilon}\ldots\overline{\epsilon}*\ldots* \ .$$
$$\vdash\!\!-m\!\!-\!\!\dashv$$

As multiplier for calculating d_{i+1} or x_{i+1} we then choose instead of

$$(2-d_i) \triangleq \overline{\epsilon}.\epsilon\ldots\epsilon\overline{*}\ldots\overline{*} \Big)$$
$$\underbrace{\phantom{\overline{\epsilon}.\epsilon\ldots\epsilon}}_{m}\phantom{\overline{*}}\underbrace{\phantom{\ldots\overline{*}}}_{+1}$$

the following representation, r digits rounded after the binary point:

$$(2-d_i)_{T_r} \triangleq \overline{\epsilon}.\epsilon\ldots\epsilon\overline{*}\ldots\overline{*}10\ldots0 \quad (m < r \leq 2m - 1).$$
$$\underbrace{\phantom{\overline{\epsilon}.\epsilon\ldots\epsilon}}_{m}$$
$$\underbrace{\phantom{\overline{\epsilon}.\epsilon\ldots\epsilon\overline{*}\ldots\overline{*}}}_{r}$$

It can now be shown that, after carrying out one iteration step with the rounded multiplier, at least the initial r digits after the point agree:

<u>Proposition 4.4.</u> $|1-d_i| \leq 2^{-m} \Rightarrow |1-d_{i+1}| = |1-d_i(2-d_i)_{T_r}| < 2^{-r}$

$$(m < r \leq 2m - 1);$$

i.e.

$$d_i \triangleq \begin{cases} 0.1\ldots.1*\ldots\ldots\ldots* \\ 1.0\ldots.0*\ldots\ldots\ldots* \\ \underbrace{}_{m} \end{cases}$$

$$(2-d_i)_{T_r} \triangleq \begin{cases} 1.0\ldots.0\overline{*}\ldots\overline{*}10\ldots.0 \\ 0.1\ldots.1\overline{*}\ldots\overline{*}10\ldots.0 \\ \underbrace{\phantom{1.0\ldots.0\overline{*}\ldots\overline{*}}}_{r} \end{cases}$$

$$d_{i+1} \triangleq \begin{cases} 0.1\ldots\ldots\ldots.1*\ldots..* \\ 1.0\ldots\ldots\ldots.0*\ldots..* \\ \underbrace{}_{r} \end{cases}$$

Hence, if d_i is exact for m digits after the point and if $(2-d_i)$ is rounded to r digits after the point, then d_{i+1} is exact for at least r digits.

<u>Proof.</u> As previously, we define $d_i := 1-x$; $2-d_i := 1+x$; we shall define:

$$x_{T_r} := (1+x)_{T_r} -1 = (2-d_i)_{T_r} -1.$$

Simple calculation indicates that the following inequalities apply:

    a. $|x| \leq 2^{-m}$;   b. $|x_{T_r}| < 2^{-m}$;   c. $|x-x_{T_r}| \leq 2^{-r-1}$.

Hence:

$$|1-d_{i+1}| = |1-(1-x)(1+x_{T_r})| = |x-x_{T_r} + x \cdot x_{T_r}|$$

$$\leq |x-x_{T_r}| + |x| \cdot |x_{T_r}| < 2^{-r-1} + 2^{-m} \cdot 2^{-m}$$

$$\leq 2^{-r-1} + 2^{-r-1} \quad \text{(since } r \leq 2m-1\text{)}$$

$$= 2^{-r} \; .$$

Selecting r=2m-1, one accuracy bit is lost at most due to imprecise calculation:

shortened multiplier: $|1-d_{i+1}| < 2^{-r} = 2^{-2m+1}$;

unshortened multiplier: $|1-d_{i+1}| < 2^{-2m}$.

The new iteration method has the following form:

Iteration method (rounded multipliers)

$y_T := f_T(b) \approx \frac{1}{b}$;

$d_1 := b \cdot y_T$; $x_1 := a \cdot y_T$;

$d_{i+1} := d_i \cdot (2-d_i)_{T_r}$; $x_{i+1} := x_i \cdot (2-d_i)_{T_r}$.

IV. Acceleration by multiplier coding

We will determine r_i in such a way as to code the shortened multiplier $(1+x)_{T_{r_i}}$ with 6 multiples respectively (coding in groups of 2+1 bits each; (see 3.3.4)).

We will assume:

$$r_i := r_{i-1} + 9 \; ;$$

i.e. $(2-d_i)_{T_{r_i}} \triangleq \begin{cases} 0.1\ldots\ldots..1*\ldots*100\ldots.0 \\ 1.0\ldots\ldots..0*\ldots*100\ldots.0 \end{cases}$

$\vdash r_{i-1} \dashv$

$\vdash\!\!\!-\!r_i\!=\!r_{i-1}\!+\!9\!\dashv$

$(2-d_i)_{T_{r_i}}$ can be coded with 7 multiples $V_0,\ldots,V_6$ in the following manner.

$\left.\begin{array}{l}1.00\ldots.0\ldots\ldots..00000**********100\\0.1111111\ldots\ldots..11111**********100\end{array}\right\}$ binary representation of $(2-d_i)_{T_{r_i}}$

$\underset{V_6=1}{\top} \qquad \underset{V_5 V_4 V_3 V_2 V_1 V_0 =-2}{\top\top\top\top\top\top}$

We can dispense with the multiple V_0. This means that instead of $(2-d_i)_{T_r}$ the multiplier $(2-d_i)_{T_r} + 2^{-r-1}$ is employed for calculating d_{i+1} or x_{i+1}. The following proposition proves that this results in no substantial loss of accuracy:

Proposition 4.5. Let $|1-d_i| \leq 2^{-m}$, and let d_{i+1} be defined as follows:

$$d_{i+1} := d_i \cdot [(2-d_i)_{T_r} + 2^{-r-1}] \qquad (m < r \leq 2m-1).$$

Then, as in the previous proposition:

$$|1-d_{i+1}| < 2^{-r}.$$

Proof. Employing the notations of proposition 4.4, we obtain:

$$\begin{aligned}|1-d_{i+1}| &= |1-(1-x)\cdot(1+x_{T_r} + 2^{-r-1})| \\ &= |(x-x_{T_r})+(x\cdot x_{T_r}-(1-x)\cdot 2^{-r-1})| \\ &\leq |x-x_{T_r}| + |x\cdot x_{T_r}-(1-x)\cdot 2^{-r-1}| \\ &\leq 2^{-r-1} + |x\cdot x_{T_r}-(1-x)\cdot 2^{-r-1}|.\end{aligned}$$

It can be shown by case distinction (separate examination of the cases x=0, x<0 and x>0) that the second addend of these inequalities can always be upvalued by 2^{-r-1}. Hence, the proposition is proved.

Result. *We can obtain at least 9 accuracy bits in the i-th iteration step by means of a shortened multiplier which can be coded with 6 multiples* $V_1,\ldots,V_6$, *if* $r_{i-1} \geq 10$.

$$d_i \triangleq \begin{cases} 0.11\ldots\ldots\ldots\ldots 1\overbrace{**************}^{r_{i-1}} \\ 1.00\ldots\ldots\ldots\ldots 0************** \end{cases}$$

where the brace spans $r_i = r_{i-1}+9$

$$(2-d_i)_{T_{r_i}} \triangleq \begin{cases} 1.00\ldots\ldots\ldots\ldots 0\overline{********}1000..0 \\ 0.11\ldots\ldots\ldots\ldots 1\overline{********}1000000 \end{cases}$$

$\underset{V_6}{\uparrow}1\underset{V_5 V_4 V_3 V_2 V_1}{\uparrow\uparrow\uparrow\uparrow\uparrow}$ *(multiplier coding)*

$$d_{i+1} = d_i \cdot [(2-d_i)_{T_{r_i}} + 2^{-r_i-1}]$$

$$\triangleq \begin{cases} 0.11\ldots\ldots\ldots\ldots\ldots\ldots 1*\ldots..* \\ 1.00\ldots\ldots\ldots\ldots\ldots\ldots 1*\ldots..* \end{cases}$$

$$\underbrace{}_{\min(2 \cdot r_{i-1}-1,\ r_{i-1}+9)}$$

$$= r_{i-1}+9 \quad (\text{since } r_{i-1} \geq 10).$$

Remark. If we have $r_{i-1} < 10$ (at the beginning of the iteration method), the rounding operations must be slightly changed; less than 9 bits can be gained from a shortened multiplier. However, the above method can be applied at latest in the next iteration step. An iteration method set up for n=56 is shown in [An2].

V. Pipelining

We have mentioned several times already that calculation of

$$d_{i+1} = d_i \cdot (2-d_i)_{T_r} \quad \text{or of}$$

$$x_{i+1} = x_i \cdot (2-d_i)_{T_r}$$

can be performed in parallel. In both cases, the same shortened multiplier is used, coded with 6 multiples; the complete length of the multiplier must be used only in the final iteration (calculation of $x_E = x_{E-1} \cdot (2-d_{E-1})$). Hence, the Wallace-tree of the multiplication in 3.6.3 can be used for all iterations (except the final).

A prime possibility for accelerating multiplication is by using two parallel Wallace-trees: fig. 4.3 illustrates the principle.

Fig. 4.3

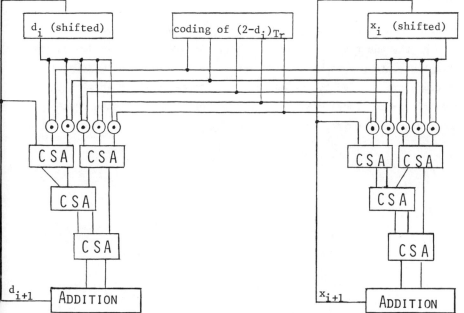

Remark. Since coding of the multiplier cannot take account of the different digit values of the multiples $V_1,\ldots,V_6$ in regard to various iterations, the multiplicand (d_i or x_i) must be shifted to the right by the respective number of digits.

The similar structure of both halves of fig. 4.3 allows the combination of both Wallace-trees to a single one and application of the pipelining concept (see fig. 4.4):

Calculation of x_{i+1} is started when the calculation process of d_{i+1} is in the lower half of the tree. If x_{i+1} is in the lower half, d_{i+2} starts in the upper half. Hence both halves of the tree are occupied in turn calculating the iteration values d_j or x_j.

The frequency of functions is chosen so that one iteration can occur in 2 functions. In order to avoid hazards, the two halves of the tree are separated by simple storage elements (latches, for example; see 3.6.2).

The final multiplication $x_E := x_{E-1}(2-d_{E-1})$ must be carried out over its full length; the result is the quotient. As previously, we divide the multiplier into groups of 12 bits each. As in 3.6.3, the multiples of one group are reduced to two addends and processed together with the next group (feed-back loop). With application of this pipelining concept, each new group requires one additional function only.

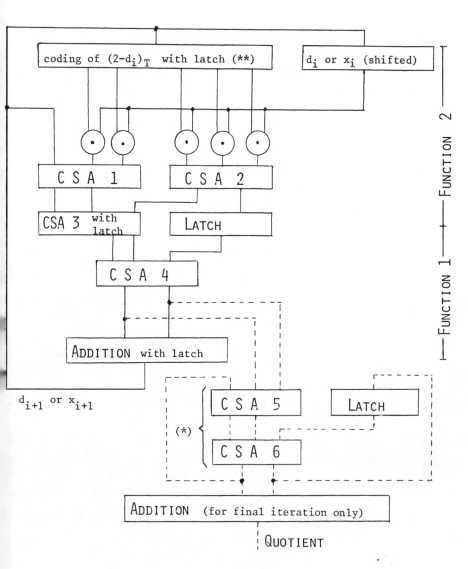

Fig. 4.4 Iterative division

(*) Feedback loop for the final iteration only: CSA 5 and CSA 6 contain latches which function individually (see 3.6.3).

(**) When coding $(2-d_i)_T$, latches are necessary which take on new values from the input lines only in every second function. Hence, d_{i+1} is taken on, after the next function x_{i+1} is blocked, then d_{i+1} is again taken on etc.

Since $|1-(2-d_{E-1})| \leq 2^{-n/2}$ (n=length of word), $(2-d_{E-1})$ can be coded with $\frac{n+8}{4}$ multiples at most. Hence the number of groups in the last multiplication amounts to $\lceil\frac{n+8}{24}\rceil$. After $\lceil\frac{n+8}{24}\rceil + 1$ functions we obtain the product $x_{E-1} \cdot (2-d_{E-1})$ in carry save form (as the sum of 2 numbers). Addition of these two addends provides the quotient.

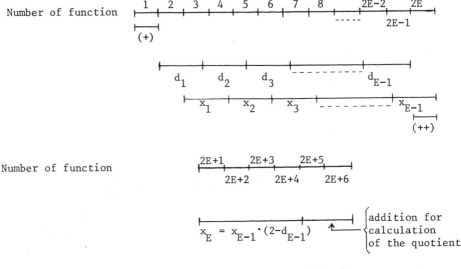

Fig. 4.5 (Iterative division: time schedule of procedure)

(+)   : Table look up for calculation of $y_T \approx \frac{1}{b}$.

(++) : Coding of first group of the multiplier$(2-d_{E-1})$.
        For n=56, this last multiplier can be coded into 3 groups of
        12 bits each (see [An2]). The final iteration (without
        final addition) requires only 4 further functions.

4.5.5 Iterative calculation of square roots [Ra2]

The principles of iterative division can be applied to calculation of
other operations without difficulty. It should be noted that in regard
to all methods, the rule of iteration must be readily implemented and
must be free of divisions.

4.5.5.1 Application of the Newton method

The Newton rule of iteration

$$x_{i+1} := \varphi(x_i) \qquad \varphi(x) := x - \frac{f(x)}{f'(x)}$$

provides square convergence towards one zero digit of $f(x)$ (see 4.5.3).
Hence, by setting:

$$f(x) := x^2 - N ,$$

the rule of iteration now reads

$$\begin{aligned} x_{i+1} &= x_i - \frac{1}{2x_i} \cdot (x_i^2 - N) \\ &= \frac{1}{2} \cdot (x_i + \frac{N}{x_i}) , \end{aligned}$$

and iteration provides a <u>square convergence towards the zero digit $\sqrt{N}$</u>
of $f(x)$; as starting value, we can choose $x_0=1$ or an approximate
value for $\sqrt{N}$.

4.5.5.2 Newton method with table look up

The iteration stated in 4.5.5.1 is not free of division and is therefore less suitable for rapid calculation of square roots. However, we can avoid division by using a table value:

$$x_{i+1} := x_i - f_T(x_i) \cdot (x_i^2 - N)$$

where $f_T(x_i)$ is a table value for $1/(2x_i)$.

The fact that the table value is less precise results in **convergence** of this rule of iteration **being only linear**.

4.5.5.3 Newton methods free of division

If $f(x) := 1 - \dfrac{1}{N \cdot x^2}$,

in the Newton method, the iteration rule reads:

$$x_{i+1} = x_i - (1 - \frac{1}{N \cdot x_i^2}) / \frac{2}{N \cdot x_i^3}$$

$$= \underbrace{\frac{x_i}{2} \cdot (3 - N \cdot x_i^2)}_{(A)} = \underbrace{\frac{1}{2} \cdot \left[(x_i + 2x_i) - (N \cdot x_i) \cdot x_i^2\right]}_{(B)}$$

Both rules (A) and (B) **provide square convergence** towards the zero digit $\frac{1}{\sqrt{N}}$ of $f(x)$. We obtain the value for $\sqrt{N}$ by multiplying by N.

For one iteration step, 3 multiplications and one addition (formation of a "3-complement") are necessary with method (A), 3 multiplications and 2 additions are necessary with method (B), while it is possible to carry out one of the multiplications/additions simultaneously with another multiplication; see the bracketed structure in (B). Hence, method (B) is faster if it is possible to carry out several arithmetic operations in parallel.

4.5.5.4 The Anderson-Earle-Goldschmidt-Powers algorithm for the calculation of square roots

The method described under 4.5.4.3 can be extended entirely to calculation of square roots, division free.

Anderson-Earle-Goldschmidt-Powers method [Ra2]

$d_0 := N \qquad\qquad x_0 := N$

$d_{i+1} := d_i \cdot r_i^2 \qquad x_{i+1} := x_i \cdot r_i$

writing $\quad r_i := 1 + \frac{1}{2}(1 - d_i)$.

It can be readily shown that:

$$d_n = d_0 \cdot r_0^2 \cdot r_1^2 \cdot \ldots \cdot r_{n-1}^2 \to 1 \quad \text{(square convergence)}$$

hence: $\quad r_0 \cdot r_1 \cdot \ldots \cdot r_{n-1} \to \dfrac{1}{\sqrt{d_0}}$.

This yields:

$$x_n = x_0 \cdot r_0 \cdot r_1 \ldots r_{n-1} \to \frac{x}{\sqrt{d_0}} = \frac{N}{\sqrt{N}} = \sqrt{N} \ .$$

As in 4.5.4.3, the d_i sequence converges towards 1 and the x_i sequence towards the desired functional value. By means of the methods described there, calculation can be accelerated or carried out with less expenditure. Details are given under [Ra2].

5. Redundant representation

5.1 SDNR representation to base $d \geq 3$

In the algorithms used so far, the duration τ of an addition is decisively influenced by the length of the largest carry propagation chain. Since the carry is passed over all digits in the most unfavourable case, τ depends on the length of the addends: if the representation is non-redundant, if only circuit elements with at most r inputs are available, and if the passing of one module requires one unit of time, the following lower limit for the maximum duration of an addition can be established (see 7.2.3):

$$\tau \geq \lceil \log_r (2n) \rceil \qquad (n = \text{length of the addends}).$$

If, however, the representation is sufficiently redundant to collect every carry at latest after a number k of bit positions which is independent from n, τ can be shortened to $\tau = O(k)$. We have to accept that this acceleration may result in higher expenditure for deconverting a redundant number into a non-redundant number.

We have already treated two redundant notations: the (d-1)-complement and the coding of a denary number by amount and sign (see 1.1); the only number with several representations is zero. It is obvious that this redundancy is not sufficient to restrict the length of the propagation chain. A simple possibility for gaining redundancies is increasing the number of digits used for the representation and also allowing negative digits. In this way, numbers can be coded in several ways (see [Av1], [Av2], [At2]).

Definition 5.1. *An example*

$$w_{red}^{(n,m)} : R \dot{\times} R \dot{\times} \ldots \ldots \dot{\times} R \to \mathbb{Q}$$

$$\alpha_{n-1} \ldots \alpha_0 \alpha_{-1} \alpha_{-2} \ldots \alpha_{-m} \longrightarrow \sum_{i=-m}^{n-1} \alpha_i \cdot d^i,$$

where $\alpha_i \in R := \{-r_1, -r_1+1, \ldots, 0, 1, \ldots, r_2-1, r_2\}$ $(r_1, r_2 > 0)$
is called <u>redundant notation with base d</u>.

This definition is too general in the form stated; hence, we make the following restrictions:

1. Let R be a symmetrical area, i.e. $r_1 = r_2 = r > 0$.
2. Each digit α_i can accept at least d different values. Furthermore, a carry (+1 or -1) resulting from position i-1 shall be collectable in position i; then carries can proceed over one position at most. Both requirements are met, if:

    $|R| = 2r+1 \geq d+2$, i.e. $\lfloor \frac{d}{2} \rfloor + 1 \leq r$.

3. α_i can accept at most d non-negative values (i.e. all denary digits), i.e. $r \leq d-1$.

This yields the following definition of redundant notation:

<u>Definition 5.2.</u> *An example*

$$w_{red}^{(n,m)} : R \dot\times R \dot\times \ldots \ldots \dot\times R \to Q$$

$$\alpha_{n-1} \ldots \alpha_0 \alpha_{-1} \ldots \alpha_{-m} \longrightarrow \sum_{i=-m}^{n-1} \alpha_i \cdot d^i ,$$

where $\alpha_i \in R := \{-r, -r+1, \ldots, 0, \ldots, r-1, r\}$ and $\lfloor \frac{d}{2} \rfloor + 1 \leq r \leq d-1$
is called <u>signed digit number representation (SDNR)</u> with base d.

If $r = \lfloor \frac{d}{2} \rfloor + 1$, the representation is called <u>minimal redundant</u>,
If $r = d-1$, the SDNR is called <u>maximal redundant</u>.

<u>Examples.</u> a. d=10, r=6   (minimal redundant)

$$w_{red}^{(5,0)} (0\ -6\ 0\ 5\ -4) = w_{red}^{(5,0)} (-1\ 4\ 0\ 5\ -4)$$
$$= w_{red}^{(5,0)} (0\ -6\ 0\ 4\ 6) = -5954.$$

b.   d=10, r=9   (maximal redundant)

$$w_{red}^{(5,0)}(0\ 7\ -8\ -8\ -2) = w_{red}^{(5,0)}(1\ -3\ -8\ -9\ 8) = 6118 \ .$$

Lemma 5.1. *(Characteristics of the SDNR):*

a. $w_{red}(\alpha_{n-1} \ldots \alpha_{-m}) = 0 \Leftrightarrow \alpha_i = 0 \ \text{for all} \ i.$

b. If
$$w_{red}(\alpha_{n-1}\ldots\alpha_{-m}) = a, \text{ then:}$$
$$w_{red}(\bar{\alpha}_{n-1}\ldots\bar{\alpha}_{-m}) = -a, \text{ where } \bar{\alpha}_i := -\alpha_i \ .$$

c. Let $k := \max\{j \mid \alpha_j \neq 0\}$.
If $w_{red}(\alpha_{n-1}\ldots\alpha_{-m}) \neq 0$, then: $\text{sign}(w_{red}(\alpha_{n-1}\ldots\alpha_{-m})) = \text{sign}(\alpha_k)$.

Lemma 5.1 shows that, contrary to residue arithmetics, (see 1.4.2), determination of the sign of a SDNR number or calculation of the representation of -a from the one of +a is very simple.

<u>Remark.</u> Since $\lfloor \frac{d}{2} \rfloor + 1 \leq r \leq d-1$, $d \geq 3$ is necessary: hence there is no binary SDNR which meets all requirements regarding the range of numbers and cycle time of the carry.

5.2 Parallel addition of SDNR addends; conversion

Redundancy of the representation allows the addition of two SDNR addends in two steps; any carries from step 1 are collected in step 2. Addition proceeds as follows:

$$\begin{array}{l}
0 \ \alpha_{n-1} \ldots \alpha_{-m+1} \alpha'_{-m} \\
0 \ \beta_{n-1} \ldots \beta_{-m+1} \beta_{-m} \\
\hline
0 \ \sigma_{n-1} \ldots \sigma_{-m+1} \sigma_{-m} \\
c_{n-1} \ c_{n-2} \ldots c_{-m} \\
\hline
s_n \ s_{n-1} \ldots s_{-m+1} s_{-m}
\end{array}
\quad \begin{array}{l} \text{step 1} \\ \\ \text{step 2} \end{array}
\qquad
\begin{array}{l}
\text{where } \alpha_i, \beta_i, s_i \in \{-r, \ldots, +r\}, \\
\sigma_i \in \{-r+1, \ldots, r-1\}, \\
c_i \in \{-1, 0, 1\}.
\end{array}$$

Since $|\sigma_i| < r$, then $|\sigma_i + c_{i-1}| \leq r$; step 2 can therefore be performed without carry propagation.

I. Formulae for the calculation of σ_i and c_i

The intermediate values σ_i and c_i are calculated as in the von Neumann addition (see 2.2.3), but we have to consider that negative digits or negative carries can also occur.

As long as the sum of digits α_i and β_i does not exceed the amount of a given threshold value w_{max}, we will place c_i on to 0, otherwise $c_i = +1$ or $c_i = -1$, i.e.

$$c_i := \begin{cases} 0 & \text{when } |\alpha_i + \beta_i| \leq w_{max} \\ 1 & \text{when } \alpha_i + \beta_i > w_{max} \\ -1 & \text{when } \alpha_i + \beta_i < -w_{max} \end{cases}$$

$$\sigma_i := \alpha_i + \beta_i - c_i \cdot d \quad (\text{i.e. } \alpha_i + \beta_i = c_i \cdot d + \sigma_i).$$

Possible values for w_{max} are restricted to the following range by the conditions $c_i \in \{-1, 0, 1\}$ and $|\sigma_i| < r$:

$$1 \leq d - r \leq w_{max} \leq r - 1.$$

If $w_{max} \geq r$, this would yield the relationship $\sigma_i = \alpha_i + \beta_i - 0 \cdot d = r$ for $\alpha_i + \beta_i = r$; but if $w_{max} \leq d - r - 1$ and $\alpha_i + \beta_i = d - r$, then $\sigma_i = d - r - 1 \cdot d = -r$ as against $|\sigma_i| < r$. Hence, w_{max} can take on values only in the range $[d-r : r-1]$. The smallest amount of carries ($c_i \neq 0$) is obtained by using the highest possible threshold value $w_{max} = r - 1$.

This addition method can also be applied for converting a denary number to a SDNR number. Here, the second addend is placed on to 0. Application of the addition formulae yields a converted denary number after two steps. Deconverting a SDNR number is simple also, since every SDNR number can be written as the difference between two denary numbers. When calculating this difference, however, carries may continue (over all digits, in the extreme case).

Result. *The cycle time for conversion does not depend on the word length n. The average time for deconversion is equal to the time necessary for the addition of two denary numbers.*

According to construction, we have:

$$w_{red}^{(n+1,m)}(0\ \alpha_{n-1}..\alpha_{-m}) + w_{red}^{(n+1,m)}(0\ \beta_{n-1}..\beta_{-m})$$
$$= w_{red}^{(n+1,m)}(0\ \sigma_{n-1}..\sigma_{-m}) + w_{red}^{(n+1,m)}(c_{n-1}..c_{-m}\ 0)$$
$$= w_{red}^{(n+1,m)}(s_n......s_{-m}).$$

Application of the SDNR is generally unsuitable for performing a single addition, if the result is to be decoded immediately.

If, however, the sum of A denary numbers of length n (A large) is to be calculated, transition to a redundant notation is recommended, since A-1 fast additions (speed is independent from the word length n) encounter only a single deconversion (after carrying out the final SDNR addition).

Example. $d=10$; $r=7$; $w_{max}=4$.

We will calculate the sum and difference of the two decimal numbers a and b.

a	≙	02890781	b	≙	05748927	conversion of the
		00000000			00000000	addends a or b
		02$\bar{2}$10$\bar{3}$21			05$\bar{3}$42$\bar{1}$23	
		01101100			11011010	
		03$\bar{1}\bar{1}$1$\bar{2}$21			1$\bar{4}$35$\bar{1}$13$\bar{3}$	

a	≙	03$\bar{1}\bar{1}$1$\bar{2}$21	a	≙	03$\bar{1}\bar{1}$1$\bar{2}$21	addition/subtraction
b	≙	1$\bar{4}$35$\bar{1}$13$\bar{3}$	-b	≙	$\bar{1}$4$\bar{3}$5$\bar{1}$1$\bar{3}$3	of a and b
		1$\bar{1}$44031$\bar{2}$			$\bar{1}$3242$\bar{1}$54	
		00000000			10$\bar{1}$00$\bar{1}$00	
a+b	≙	1$\bar{1}$44031$\bar{2}$	a-b	≙	03$\bar{1}$42$\bar{2}$54	

		10040010			00142054	deconversion of
		-01400302			-03000200	the result
a+b	≙	08639708	a-b	≙	-02858146	

5.3 Application of SDNR numbers regarding multiplication/division

Most multiplication methods consist of a series of additions of denary numbers as well as shifts.

Additions can be expedited by transition to SDNR numbers. (One shift of a SDNR number does not differ from one shift of a denary number). Parallel multiplication methods and multiplier coding can also be applied to SDNR without difficulty. Analog, SDNR numbers can be used in division methods.

An argument against the practical application of SDNRs is the fact that there are no SDNR numbers with base d=2 and that recognition and correction of overflow is quite tedious. We will discuss below a method allowing the inclusion of binary numbers in SDNR addition arithmetics.

Finally, we will deal also with a division method which calculates the quotient of 2 denary operands in the form of an SDNR number (SRT division).

5.4 Parallel addition/subtraction with different representation of the addends

5.4.1 Addition modulo d^n

We will describe a method for adding two n-digit numbers in two steps; it is assumed that the first addend is a maximum redundant SDNR number with base d (i.e. $R = \{-d+1,\ldots,d-1\}$), and that the second is a denary number represented in the d-complement.

The result of the addition is again a maximum redundant SDNR number. The addition process is obtained by generalisation of the method discussed in 5.2 (lower case letters are used for denary and upper case for SDNR digits):

Diagram of parallel addition/subtraction, modulo d^n		
$A \triangleq A_{n-1}\ldots A_1 A_0$	$A \triangleq A_{n-1}\ldots A_1 A_0$	$A_i \in \{-d+1,\ldots,d-1\}$;
$+b \triangleq \underline{b_{n-1}\ldots b_1 b_0}$	$-b \triangleq \underline{b^*_{n-1}\ldots b^*_1 b^*_0}$	$b_i \in \{0,\ldots\ldots,d-1\}$;
$T_{n-1}\ldots T_1 T_0$	$T_{n-1}\ldots T_1 T_0$	$b^*_i := d-1-b_i$
$\underline{c_{n-1} c_{n-2}\ldots c_0 0}$	$\underline{c_{n-1} c_{n-2}\ldots c_0 1}$	$\in \{0,\ldots\ldots,d-1\}$.
$(S_n) S_{n-1}\ldots S_1 S_0$	$(S_n) S_{n-1}\ldots S_1 S_0$	

The values c_i, T_i, S_i are calculated as follows (see 5.2):

1. $c_i = \begin{cases} 1 & \text{where } A_i + b_i \geq d-1 \\ 0 & \text{otherwise.} \end{cases}$

2. $T_i = A_i + b_i - c_i \cdot d$;   i.e. $T_i \in \{-d+1,\ldots,d-2\}$.

3. $S_i = T_i + c_{i-1}$;   i.e. $S_i \in \{-d+1,\ldots,d-2,d-1\}$.

It is apparent that all c_i and all T_i values can be calculated in parallel (first step of addition). Since $T_i \neq d-1$ and $c_i \in \{0,1\}$, carries no longer occur when calculating S_i (second step).

Subtraction A-b is based on addition A+(-b). In this case, all bits b_i must be replaced by $d-1-b_i$; c_{-1} is set at 1 (d-complement notation); c_{-1} has the value 0 for an addition. Thus, addition of a 1 when calculating the representation of -b from that of +b does not result in lost time.

Correctness of the result:

As demonstrated below, the result of addition/subtraction is correct only to terms of order d^n:

$$w_{red}^{(n)} (A_{n-1} \ldots A_0) + w_d^{(n)} (b_{n-1} \ldots b_0)$$

$$= w_{red}^{(n)} (A_{n-1} \ldots A_0) + w_{red}^{(n)} (-\text{sign}(b_{n-1}) \; b_{n-2} \ldots b_0)$$

$$\overset{(*)}{=} w_{red}^{(n+1)} (0 \; T_{n-1} \ldots T_0) + w_{red}^{(n+1)} (c_{n-1} \ldots c_0 c_{-1}) \bmod d^n$$

$$= w_{red}^{(n+1)} (S_n S_{n-1} \ldots S_0) \bmod d^n$$

$$= w_{red}^{(n)} (S_{n-1} \ldots S_0) \bmod d^n$$

The validity of the equation (*) is shown by simple consideration of the calculation of T_{n-1} and c_{n-1} for the two alternatives $b_{n-1}=0$ and $b_{n-1}=d-1$.

Obviously S_n and c_{n-1} do not need to be calculated since deviations of the order d^n can occur; see examples below:

Examples. $d=10$, $n=5$.

1. $A = 009\bar{3}\bar{5}$
    $b = \underline{96123}$
    $\bar{1}60\bar{1}\bar{2}$
    $\underline{101000}$
    $1\bar{1}70\bar{1}\bar{2}$

2. $A = \bar{9}2794$
    $b = \underline{96123}$
    08817
    $\underline{000100}$
    008917

3. $A = 996\bar{3}2$
    $b = \underline{02416}$
    $\bar{1}\bar{1}2\bar{4}8$
    $\underline{110000}$
    $101\bar{2}48$

On the basis of the relationship

$$w_d^{(n)}(b_{n-1}\ldots b_0) = w_{red}^{(n)}(-\text{sign}(b_{n-1})\, b_{n-2}\ldots b_0)$$

calculations would proceed as follows with transition to a redundant representation of the addend b:

1. $A = 009\bar{3}\bar{5}$
    $b_{red} = \underline{\bar{1}6123}$
    $\phantom{b_{red} = }\bar{1}60\bar{1}\bar{2}$
    $\phantom{b_{red} = }\underline{001000}$
    $\phantom{b_{red} = }0\bar{1}70\bar{1}\bar{2}$

2. $A = \bar{9}2794$
    $b_{red} = \underline{\bar{1}6123}$
    $\phantom{b_{red} = }08817$
    $\phantom{b_{red} = }\underline{\bar{1}00100}$
    $\phantom{b_{red} = }\bar{1}08917$

3. (as above)

In all cases, the result corresponds to the correct value except for a term of the order 10^5.

It is obvious that application of this form of SDNR addition depends mainly on the existence of a simple overflow criterion.

5.4.2 Overflow recognition/correction

As in 1.3, overflow can be recognized by extension of the notation by one digit; furthermore, redundancy of representation provides an automatic overflow correction in most cases. One position $A_n = 0$ must be added to the SDNR addend, and the respective position of the addend represented in the d-complement is obtained by doubling of the sign, as in 1.3.

Adder diagram with overflow recognition

$$\begin{array}{l} \phantom{(S_{n+1})\ }0\ \ A_{n-1}A_{n-2}\cdots\cdots A_1 A_0 \\ \phantom{(S_{n+1})\ S_n S_{n-1}}\underline{u_{n-1}u_{n-1}u_{n-2}\cdots\cdots u_1 u_0} \\ \phantom{(S_{n+1})\ }T_n\ T_{n-1}T_{n-2}\cdots\cdots T_1 T_0 \\ \phantom{(S_{n+1})\ }\underline{(c_n)c_{n-1}c_{n-2}c_{n-3}\cdots\cdots c_0 c_{-1}} \\ (S_{n+1})\ S_n\ S_{n-1}S_{n-2}\cdots\cdots S_1 S_0 \end{array}$$

where $A_i, S_i \in \{-d+1,\ldots,d-1\}$;

$T_i \in \{-d+1,\ldots,d-2\}$;

$$u_i := \begin{cases} b_i & \text{addition } A+b \\ d-1-b_i & \text{subtraction } A-b \end{cases}$$

$$c_{-1} := \begin{cases} 0 & \text{addition } A+b \\ 1 & \text{subtraction } A-b \end{cases}$$

Overflow can now be treated as follows:

Lemma 5.2. *a. There is precisely an overflow, if $S_n \neq 0$; position S_{n+1} does not need to be calculated.*

b. The overflow can be corrected if the two first digits from the left in $S_n S_{n-1}\ldots S_0$, which differ from zero, have different signs.

If the sum has the form

$$S_n S_{n-1}\ldots S_0 = S_n \underbrace{0\ldots\ldots 0}_{k \geq 0} S_{n-k-1}S_{n-k-2}\ldots S_0 ,$$

where $S_n, S_{n-k-1} \neq 0$; $\text{sign}(S_n) \neq \text{sign}(S_{n-k-1})$,

this overflow situation can be corrected to

$$0 S'_{n-1}\ldots S'_0$$

with

$$S'_i := \begin{cases} d-1 & i = n-1,\ldots,n-k \\ d+S_i & i = n-k-1 \\ S_i & i = n-k-2,\ldots,0 \end{cases} \quad \text{when} \quad S_n = 1, S_{n-k-1} < 0$$

or $\quad S'_i := \begin{cases} \overline{d-1} & i = n-1,\ldots,n-k \\ \overline{d}+S_i & i = n-k-1 \\ S_i & i = n-k-2,\ldots,0 \end{cases}$ when $\quad S_n = \overline{1}, S_{n-k-1} > 0$.

Proof. Since

$$w_d(u_{n-1}u_{n-1}u_{n-2}\cdots u_0) = w_{red}(-\text{sign}(u_{n-1})\ u_{n-1}u_{n-2}\cdots u_0)$$

we have the following addition procedure due to transition to redundant representation for the addend represented in the d-complement:

1. $u_{n-1} = 0$

$$\begin{array}{l} 0\ A_{n-1}A_{n-2}\cdots A_0 \\ \underline{0\ 0\ u_{n-2}\cdots u_0} \\ 0\ldots\ldots\ldots\ldots \\ \underline{c_{n-1}\cdots\cdots\cdots\cdots} \\ S_n\ S_{n-1}\cdots\cdots\cdots \end{array}$$

$S_n = c_{n-1} = \begin{cases} 1 \Leftrightarrow A_{n-1} = d-1 \\ 0 \text{ otherwise} \end{cases}$

2. $u_{n-1} = d-1$

$$\begin{array}{l} 0\ A_{n-1}A_{n-2}\cdots A_0 \\ \underline{\overline{1}\ d-1\ u_{n-2}\cdots u_0} \\ \overline{1}\ldots\ldots\ldots\ldots \\ \underline{c_{n-1}\cdots\cdots\cdots\cdots} \\ S_n\ S_{n-1}\cdots\cdots\cdots \end{array}$$

$S_n = c_{n-1} + \overline{1} = \begin{cases} \overline{1} \Leftrightarrow A_{n-1} < 0 \\ 0 \text{ otherwise} \end{cases}$

This immediately yields:

$$w_{red}^{(n+1)}(0\ A_{n-1}\cdots A_0) + w_d^{(n+1)}(u_{n-1}u_{n-1}u_{n-2}\cdots u_0)$$
$$= w_{red}^{(n+1)}(S_n S_{n-1}\cdots S_0) = w_{red}^{(n)}(S_{n-1}\cdots S_0) \bmod d^n .$$

Hence, an overflow situation is indicated by $S_n \ne 0$; it can be corrected if the number can be represented with n digits, i.e., if the next position different from zero in $S_{n-1}\cdots S_0$ has a sign different from S_n.

Automatic correction of <u>all</u> overflow situations is performed at the expense either of cycle time or expenditure. Hence, we will be content with a correction of those overflow situations where length k of the zero block following S_n does not exceed the fixed value k*. In most cases we restrict ourselves to k*=0, i.e. to comparison of the two most significant bits of the intermediate sum $S_n...S_0$ (see [Av2]); this already covers the majority of overflow situations which can be corrected; proposition 5.3 demonstrates this:

Proposition 5.3. *(Automatic overflow correction for k*=0)*

$$0 \quad A_{n-1}A_{n-2}...A_1A_0$$
$$u_{n-1}u_{n-1}u_{n-2}...u_1u_0$$
$$T_n \, T_{n-1}T_{n-2}...T_1T_0$$
$$c_{n-1}c_{n-2}c_{n-3}...c_0c_{-1}$$
$$S_n \, S_{n-1}S_{n-2}...S_1S_0$$
$$\varepsilon_n \, \varepsilon_{n-1}$$
$$\overline{\quad\quad\quad\quad\quad\quad\quad}$$
$$S'_n \, S'_{n-1}S'_{n-2}...S'_1S'_0$$

A_i, u_i, T_i, c_i, S_i as previously

$$\varepsilon_n := \begin{cases} 0 & \text{when } S_n = 0 \text{ or } S_{n-1} = 0 \\ \overline{S_n} & \text{otherwise.} \end{cases}$$

$$\varepsilon_{n-1} := -d \cdot \varepsilon_n = \begin{cases} 0 & \text{when } S_n = 0 \text{ or } S_{n-1} = 0 \\ d \cdot S_n & \text{otherwise.} \end{cases}$$

a. *An overflow remains uncorrected only if $S'_n \neq 0$.*

b. $S'_n \neq 0 \Leftrightarrow A_{n-1} = d-1, c_{n-2} = 1, u_{n-1} = 0 \text{ or } A_{n-1} = \overline{d-1}, c_{n-2} = 0, u_{n-1} = d-1$.

If the same probability applies to all digits A_i and u_i and if they are not interdependent, we have:

$$P(S'_n \neq 0) = \frac{1}{4d-2}.$$

Approximately each second remaining overflow situation ($S'_n \neq 0$) could be corrected; this is not considered in the adding diagram shown.

Proof. We need only to check b. Due to distinction of cases, we have:

$S_n' \neq 0 \Leftrightarrow S_n = 1 \wedge S_{n-1} \geq 0 \vee S_n = \bar{1} \wedge S_{n-1} \leq 0$

$\Leftrightarrow A_{n-1} = d-1, c_{n-2} = 1, u_{n-1} = 0$ or $A_{n-1} = \overline{d-1}, c_{n-2} = 0, u_{n-1} = d-1$.

The conditions for the digits A_i yield:

$$P(S_n' \neq 0) = \frac{1}{2d-1} \cdot \frac{1}{2} \cdot P(c_{n-2}=1) + \frac{1}{2d-1} \cdot \frac{1}{2} \cdot P(c_{n-2}=0)$$

$$= \frac{1}{4d-2} \cdot [P(c_{n-2}=1) + P(c_{n-2}=0)] = \frac{1}{4d-2} .$$

Examples. (d = 10, n = 5)

[1]	A = 78615	$08\bar{2}7\bar{9}5$		[2]	A = -82273	$09\bar{8}2\bar{8}7$
	b = -9842	990158			b = -8291	991709
	A + b = ?	$\bar{1}7\bar{2}8\bar{4}3$			A + b = ?	$\bar{1}0\bar{1}5\bar{8}6$
		✗100010				✗010010
	68773	$07\bar{2}8\bar{3}3$				$\bar{1}\bar{1}\bar{1}5\bar{7}6$ } overflow
						$1\bar{d}$ } correction
					-90564	$0\bar{9}1\bar{5}7\bar{6}$

[3]	A = 80658	$09\bar{9}4\bar{6}\bar{2}$		[4]	A = 78615	$08\bar{2}7\bar{9}5$
	b = +1398	001398			b = -9842	990158
	A + b = ?	$01\bar{8}1\bar{5}6$			A − b = ?	$08\bar{2}7\bar{9}5$
		100100				009841
		$1\bar{1}8056$ } overflow				$087\bar{5}56$
		$\bar{1}\bar{d}$ } correction				001001
	82056	$09\bar{8}056$				$088\bar{5}57$
						00
					88457	$088\bar{5}57$

5	A = -95776	0$\bar{9}$63$\bar{8}$4
b = -7125	992875
A + b = ?	$\bar{1}$041$\bar{1}\bar{1}$
	⟨001010
-102901	$\bar{1}$03$\bar{1}$0$\bar{1}$

non-correctable overflow

6	A = -88354	0$\bar{9}$2354
b = -3877	996123
A + b = ?	$\bar{1}$082$\bar{3}$1
	⟨000000
-92331	$\bar{1}$082$\bar{3}$1
	0$\bar{9}$22$\bar{3}$1

overflow correction with $k^* \geq 1$.

5.4.3 Application to a binary base [At2]

Although there are no SDNR numbers with base $d = 2$, the addition method discussed in 5.4.1 can be performed also for binary numbers. This provides fast calculation of a sum consisting of numerous binary addends. Only the time for deconverting the results depends on the length n of the addends.

We have: $A_i, S_i, S'_i \in \{-1, 0, 1\}$; $T_i \in \{-1, 0\}$; $c_i \in \{0, 1\}$.

The digits A_i, S_i, S'_i must be coded with two binary positions, while one binary position is sufficient for the remainder. We shall choose representation by amount and sign as coding (see [At2]).

Table 5.1.
coding of $A_i = (a_{i,1}, a_{i,2})$

$a_{i,1}$	$a_{i,2}$	A_i
0	0	+0
0	1	+1
1	0	-0
1	1	-1

Results (cf. the diagram in 5.4.1.):

1. $A_i = a_{i,2} \cdot (1 - 2a_{i,1})$, i.e. $a_{i,2} = A_i \mod 2$.

2. $c_i = 1 \Leftrightarrow A_i + b_i \geq 1 \Leftrightarrow A_i = 1 \vee A_i = 0 \wedge b_i = 1$,

   i.e. $c_i = \overline{a_{i,1}} \cdot a_{i,2} \vee \overline{a_{i,2}} \cdot b_i$.

3. $T_i = -1 \Leftrightarrow A_i + b_i = 1 \mod 2 \Leftrightarrow a_{i,2} + b_i = 1 \mod 2$,

   i.e. $T_i = -(a_{i,2} \oplus b_i)$; $|T_i| = a_{i,2} \oplus b_i$.

The formulae for coding $S_i = (s_{i,1}, s_{i,2})$ can also be stated immediately. We have $S_i = T_i + c_{i-1}$, thus:

4. $s_{i,2} = 1 \Leftrightarrow S_i \neq 0 \Leftrightarrow T_i + c_{i-1} = 1 \mod 2 \Leftrightarrow |T_i| \oplus c_{i-1} = 1$,

   i.e. $s_{i,2} = |T_i| \oplus c_{i-1} = a_{i,2} \oplus b_i \oplus c_{i-1}$.

5. Because of redundancy of representation (S = 0 could be coded by (0,0) or by (1,0), see table 5.1), there are several possibilities for $s_{i,1}$, for example:

$$s_{i,1} = |T_i| \quad \text{or} \quad s_{i,1} = \overline{c_{i-1}}.$$

The digits S'_n and S'_{n-1} also can be coded in a different manner. Table 5.2 indicates the simplest procedures (see table 5.1 for the coding rules):

$s_{n,1}$	$s_{n,2}$	$s_{n-1,1}$	$s_{n-1,2}$	$s'_{n,1}$	$s'_{n,2}$	$s'_{n-1,1}$	$s'_{n-1,2}$
α	1	β	0	α	1	*	0
α	1	$\overline{\alpha}$	1	*	0	α	1
α	1	α	1	α	1	α	1
α	0	β	γ	*	0	β	γ

$s'_n = s_n + \varepsilon_n$

$s'_{n-1} = s_{n-1} + \varepsilon_{n-1}$

Table 5.2

($\alpha, \beta, \gamma \in B$; the position * can take 0 or 1 at will).

We obtain: $s'_{n,1} = s_{n,1}$; $s'_{n-1,2} = s_{n-1,2}$;

$s'_{n,2} = s_{n,2} \cdot [\overline{s_{n-1,2}} \vee (\overline{s_{n-1,1} \oplus s_{n,1}})]$;

$s'_{n-1,1} = s_{n,1} \cdot s_{n,2} \vee \overline{s_{n,2}} \cdot s_{n-1,1}$.

The following example demonstrates application of these formulae for calculating a sum consisting of several addends. It is evident that the notation chosen is highly suitable for such problems (one addend is binary coded, the second and the result are SDNR numbers).

Example: (n = 7, doubling of sign)

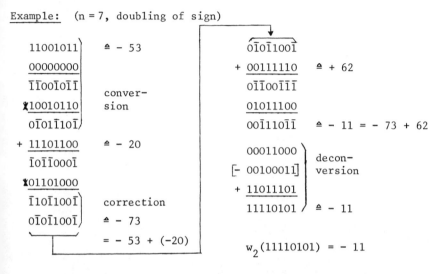

5.5 SRT division (Sweeney, Robertson [Ro1], Tocher [To1])

5.5.1 Motivation

The SRT division calculates the quotient of two denary numbers in the form of an SDNR number. The same recursion formula is used as in the restoring division (see 4.2.1):

$$X^{(n)} := w(DD,DE) \qquad w(DD,DE) \geq 0,$$
$$X^{(j)} := d \cdot [X^{(j+1)} - q_j \cdot w(DR)] \quad (j=n-1,\ldots,0) \qquad w(DR) \geq 0.$$

While only non-negative quotient bits were allowed in the methods discussed so far, an attempt is now made to make the new partial remainder $X^{(j)}$ in the SRT division (Sweeney, Robertson and Tocher) as small as possible as regards quantity by allowing negative quotient bits q_j. Hence, the average number of quotient bits which can be determined per division cycle increases together with the speed of division. This method is of special advantage when continuing to operate with the redundant representation of the quotient Even if the quotient is required in denary form (deconversion by subtracting two denary numbers, see 5.2), the method saves time in most cases because of the higher average size of shifts. We will now describe a variant of the SRT division which deviates slightly from the original method ([Ro1], [To1], [Fr1]), but which can be better integrated into the register configuration chosen by us.

5.5.2 Formulae of the denary SRT division

Let dividend and divisor be stored in the registers (DD,DE) and DR; we shall apply the d-complement for representation of negative numbers. Before starting the calculation, the operands are normalized, so that:

$$0 \leq |w(DD,DE)| < w(DR) \quad \text{and} \quad \frac{1}{d} \leq |w(DR)| < 1 \quad (\text{see } 4.1).$$

Recursion formula of the SRT division

$$X^{(n)} := w(DD,DE) \;;$$
$$X^{(j)} := X^{(j)}(q_j) := d \cdot [X^{(j+1)} - q_j \cdot w(DR)] \qquad (j=n-1,\ldots,0) \;,$$

while q_j is chosen so that:

a. $q_j \in R := \{-r, -r+1, \ldots, -1, 0, 1, \ldots, r-1, r\}$ $\quad (\lceil \frac{d}{2} \rceil \leq r \leq d-1)$

b. $|X^{(j)}(q_j)| \leq |X^{(j)}(q_j')|$ for all $q_j' \in R$.

Remark. a. Range R differs slightly from the range quoted in definition 5.2 ($\lceil \frac{d}{2} \rceil \leq r \leq d-1$ instead of $\lfloor \frac{d}{2} \rfloor + 1 \leq r \leq d-1$); this allows application of the formulae with binary base. Extension of range R is permissible since the problem of restricting the cycle time of the carry (see 5.1) does not occur as regards the SRT division.

b. In general, q_j is not yet uniquely defined by the conditions stated.

A quotient bit q_j which meets the requirements can be determined by adding $w(DR)$ to $X^{(j+1)}$ or subtracting it from $X^{(j+1)}$ until:

$$|X^{(j+1)} \pm m \cdot w(DR)| \le \tfrac{1}{2} \cdot |w(DR)| \ .$$

Following these m operations, we obtain:

$$|X^{(j)}| = d \cdot |X^{(j+1)} \pm m \cdot w(DR)| \le \tfrac{d}{2} \cdot |w(DR)|$$

and a permissible quotient bit q_j:

$$q_j := \begin{cases} +m & \text{when} \quad \text{sign}(X^{(j+1)}) = \text{sign}(w(DR)) \\ -m & \text{when} \quad \text{sign}(X^{(j+1)}) \ne \text{sign}(w(DR)). \end{cases}$$

Reason. $w(DR)$ is subtracted m times (i.e. $q_j = +m$), if $X^{(j+1)}$ and $w(DR)$ have the same sign, otherwise $w(DR)$ is added m times ($q_j = -m$).

5.5.3 Binary SRT division

In the binary case, the formulae for determining the quotient bits are especially simple (there is one addition/subtraction at most). For $d = 2$, 5.5.2 yields:

(SRT 1)
$$\begin{cases} q_j := \begin{cases} +\text{sign}(w(DR)) & \text{when} \quad X^{(j+1)} > \tfrac{1}{2}|w(DR)| \\ 0 & \text{when} \quad -\tfrac{1}{2}|w(DR)| \le X^{(j+1)} \le \tfrac{1}{2}|w(DR)| \\ -\text{sign}(w(DR)) & \text{when} \quad X^{(j+1)} < -\tfrac{1}{2}|w(DR)|; \end{cases} \\ X^{(j)} := 2 \cdot [X^{(j+1)} - q_j \cdot w(DR)]. \end{cases}$$

Determination of the quotient bit $q_j \in \{-1, 0, 1\}$ is based on comparison of two binary numbers, and is therefore relatively tedious. Hence, Sweeney, Robertson and Tocher have proposed the following simpler criterion for determining the quotient bits:

$$\text{(SRT 2)} \quad q_j := \begin{cases} +\text{sign}(w(DR)) & \text{when} \quad X^{(j+1)} \geq \frac{1}{2} \\ 0 & \text{when} \quad -\frac{1}{2} \overset{(<)}{\leq} X^{(j+1)} < \frac{1}{2} \\ -\text{sign}(w(DR)) & \text{when} \quad X^{(j+1)} \overset{(\leq)}{<} -\frac{1}{2} \end{cases}$$

The signs in brackets apply when representing $X^{(j+1)}$ in the 1-complement, those without brackets apply when using the 2-complement.

SRT division (hereinafter called (SRT 2)) can be applied for all notations of dividend and divisor. By means of the 1- or 2-complement notation, we obtain specially simple criteria for quotient bits.

If $\alpha_{n-1}^{(j+1)}$ and $\alpha_{n-2}^{(j+1)}$ are the two leading bits of the binary representation of $X^{(j+1)}$, then:

$$\text{(SRT 2)} \quad q_j = \begin{cases} +\text{sign}(w(DR)) & \text{when} \quad \alpha_{n-1}^{(j+1)} = 0 \wedge \alpha_{n-2}^{(j+1)} = 1 \\ 0 & \text{when} \quad \alpha_{n-1}^{(j+1)} = \alpha_{n-2}^{(j+1)} \\ -\text{sign}(w(DR)) & \text{when} \quad \alpha_{n-1}^{(j+1)} = 1 \wedge \alpha_{n-2}^{(j+1)} = 0 \end{cases} ;$$

i.e. $\quad q_j = (-\alpha_{n-1}^{(j+1)} + \alpha_{n-2}^{(j+1)}) \cdot \text{sign}(w(DR))$.

Simultaneous determination of several quotient bits (by shifts)

According to the construction, we have for (SRT 2):

$$q_j = 0 \Leftrightarrow \alpha_{n-1}^{(j+1)} = \alpha_{n-2}^{(j+1)} .$$

Assuming that the binary representation has the following form:

$$X^{(j+1)} \triangleq \begin{cases} 0.0\ldots0\ 1**\ldots\ldots \\ 1.1\ldots1\ 0**\ldots\ldots \end{cases} .$$
$$\vdash k \geq 1 \dashv$$

Then we can determine k quotient bits simultaneously, applying the criteria given by (SRT 2):

$$q_j = q_{j-1} = \ldots = q_{j-k+1} = 0$$

(simultaneous shift to the left of the partial remainder over k positions).

The next quotient bit q_{j-k} differs from zero. Note that the quotient bits regarding the SRT division do not depend on whether there are shifts over a leading block of zeros or ones (contrary to 4.2.4).

Treatment of the division remainder

The division remainder $R = X^{(0)}/2$, arising from the (SRT) division does not always have the same sign as the quotient.

We can bring about

sign(R) = sign(quotient)   by correction of remainder or quotient (see 4.2 - 4.4). We shall not do this here since the remainder before such correction is usually smaller than afterwards.

5.5.4   Microprograms and examples

The dividend is stored in the double length register (DD,DE). The second half is used for taking quotient bits. Since the quotient bits can have three different values, we must provide two register positions per bit. We will therefore use a second register DH and store the positive quotient bits in DE, and the negative ones in DH. Deconversion is carried out by subtraction of these two registers at the end of the microprogram. The partial remainders and division remainder are independent from the sign of the divisor (see formulae in 5.5.3); hence, the division remainder has the wrong sign if the divisor is negative; it is corrected simultaneously with deconversion of the quotient.

Microprogram S1   [Calculation of quotient bits according to (SRT 2)]

0 : (DD,DE) := dividend; DR := divisor; Z := n; DH := 0;

1 : $\underline{if}$ $DR_{n-1} = DR_{n-2}$ $\underline{then}$ [SHL(DD,DE); $DE_0:=0$; SHL(DR); $DR_0:=0$; $\underline{goto}$ 1];

1* : $\underline{if}$ DR = 10...0 $\underline{then}$ [(DE,DD) := $(\overline{DD,DE})$ + 0....1 ; $\underline{goto}$ 5];

2 : Z := Z-1; q := $-DD_{n-1} + DD_{n-2}$;

  $\underline{if}$ $DD_{n-1} \neq DD_{n-2}$ $\underline{then}$ [$\underline{if}$ $DD_{n-1} \neq DR_{n-1}$ $\underline{then}$ DD := DD + DR
                     $\underline{else}$ DD := DD - DR];

3 : (DE_0, DH_0) := $\underline{if}$ q = +1 $\underline{then}$ $(\overline{DR_{n-1}}, DR_{n-1})$

      $\underline{else}$ $\underline{if}$ q = -1 $\underline{then}$ $(DR_{n-1}, \overline{DR_{n-1}})$ $\underline{else}$ (0,0);

  $\underline{if}$ Z > 0 $\underline{then}$ [SHL(DD,DE); SHL(DH); $\underline{goto}$ 2] $\underline{else}$ [SHL(DE), SHL(DH)];

4 : DE := DE - DH; $\underline{if}$ DR_{n-1} = 1 $\underline{then}$ DD := $\overline{DD}$ + 0...01 ;

5 : END.

Remark. For understanding function 1*, compare microprogram D4 (see 4.3). The following microprogram is an extension of S1 (simultaneous calculation of several quotient bits by shifting over leading zeros/ones of the partial remainder):

Microprogram S2   [(SRT 2) with shifts over zeros and ones]

0 : (DD,DE) := dividend; DR := divisor; Z := n; DH := 0;

1 : $\underline{if}$ $DR_{n-1} = DR_{n-2}$ $\underline{then}$ [SHL(DD,DE); $DE_0:=0$; SHL(DR); $DR_0:=0$; $\underline{goto}$ 1];

1* : $\underline{if}$ DR = 10...0 $\underline{then}$ [(DE,DD) := $(\overline{DD,DE})$ + 0....1 ; $\underline{goto}$ 6];

2 : q := $-DD_{n-1} + DD_{n-2}$;

  $\underline{if}$ $DD_{n-1} \neq DD_{n-2}$ $\underline{then}$ [$\underline{if}$ $DD_{n-1} \neq DR_{n-1}$ $\underline{then}$ DD := DD + DR
                       $\underline{else}$ DD := DD - DR];

215

3 : $m := \min(k,Z)$; <here, k+1 is the length of the leading zero/one block in DD>

4 : $(DE_{m-1}, DH_{m-1}) := \underline{if}\ q = +1\ \underline{then}\ (\overline{DR_{n-1}}, DR_{n-1})$
      $\underline{else\ if}\ q = -1\ \underline{then}\ (DR_{n-1}, \overline{DR_{n-1}})\ \underline{else}\ (0,0)$;

   $(DE_i, DH_i) := (0,0) \quad (i = 0,\ldots,m-2)$;

   SHL(DH) over m positions;

   if $m \neq Z\ \underline{then}\ [Z := Z-m;\ \text{SHL(DD,DE) over m positions};\ \underline{goto}\ 2]$
      $\underline{else}\ [Z := 0;\ \text{SHL(DD,DE) over m positions, last shift over DE only}]$;

5 : $DE := DE - DH$; $\underline{if}\ DR_{n-1} = 1\ \underline{then}\ DD := \overline{DD} + 0\ldots\ldots 1$;

6 : END.

Examples of SRT division (Microprogram S2)

1) $(DD,DE) = 00100010010100 \qquad DR = 1000011 \qquad -DR = 0111101$

   $w(DD,DE) = \dfrac{1098}{2^{12}} \qquad w(DR) = \dfrac{61}{2^6} \qquad \dfrac{w(DD,DE)}{w(DR)} = -\dfrac{18}{2^6}$.

```
            DD      DE
         0010001  0010100        q := 0
         0100010  0101000        q := +1
         1000011         0
         1100101
         1001010  1010000  DE    q := -1
         0111101        01  DH
         0000111
         0111101  0000100        q := +1
         1000011        01000
         0000000
         0000000  0010000 ⎫
correc- ⎛1111111  0100010 ⎬  DE - DH
tion of ⎜     +1  1101110 ⎭
remain- ⎝0000000
der               quotient
```

2) (DD,DE) = 10010011100110   DR = 1000111   −DR = 0111001

$w(DD,DE) = -\dfrac{3469}{2^{12}}$ $w(DR) = -\dfrac{57}{2^6}$ $\dfrac{w(DD,DE)}{w(DR)} = +\dfrac{61}{2^6} - \dfrac{8}{57} \cdot \dfrac{1}{2^6}.$

	DD	DE	
	1001001	1100110	q := −1
	0111001		
	0000010		
	0101100	1101000 ← DE	q := +1
	1000111	0000 ← DH	
	1110011		
	1001111	0100000	q := −1
	0111001	000010	
	0001000		
	0001000	1000001 ⎫	
correction of	1110111	0000100 ⎭	DE − DH
remainder, since	+1		
w(DR) < 0	1111000	0111101 ← quotient	
	remainder		

3) (DD,DE) = 11010101000100   DR = 0100010   −DR = 1011110

$w(DD,DE) = -\dfrac{1374}{2^{12}}$ $w(DR) = \dfrac{34}{2^6}$ $\dfrac{w(DD,DE)}{w(DR)} = -\dfrac{40}{2^6} - \dfrac{14}{40} \cdot \dfrac{1}{2^6}.$

	DD	DE		
	1101010	1000100		q := 0
	1010101	0001000	DE	q := −1
	0100010		DH	
	1110111			
	1011100	0100000		q := −1
	0100010	010		
	1111110			
	1110010	0000000 ⎫		
		0101000 ⎭	DE − DH	
	remainder	1011000 ← quotient		

4) (DD,DE) = 01000010111010   DR = 1001110   −DR = 0110010

$w(DD,DE) = \dfrac{2141}{2^{12}}$ $w(DR) = -\dfrac{50}{2^6}$ $\dfrac{w(DD,DE)}{w(DR)} = -\dfrac{43}{2^6} + \dfrac{9}{50} \cdot \dfrac{1}{2^6}.$

```
                    DD       |  DE
                   ─────────  ──────────
                   0100001   | 0111010              q := +1
                   1001110   |
                   ───────   |
                   1101111   |
                   1011110   | 1110100    DE        q := -1
                   0110010   |    1       DH
                   ───────   |
                   0010000   |
                   0100001   | 1101001              q := +1
                   1001110   |     10
                   ───────   |
                   1101111   |
                   1011111   | 1010010              q := -1
                   0110010   |    101
                   ───────   |
                   0010001   |
                   0100011   | 0100101              q := +1
                   1001110   |    1010
                   ───────   |
                   1110001   |
                   1000101   | 0010100              q := -1
                   0110010   |  101010
                   ───────   |
                   1110111   | 0101001  ⎫
                 ⎛ 0001000   | 1010100  ⎬  DE - DH
correction of    ⎜     +1    | 1010101  ⎭
remainder        ⎝ 0001001   |──────────
                             | quotient
```

6. Calculation of special functions

The arithmetic unit of a computer usually contains only operations for the 4 basic types of calculation. More complicated arithmetic operations (for example trigonometric functions, logarithms, roots etc.) are performed by sub-programs; one exception is, for example, the calculation of square roots by the iteration method discussed in 4.5.5. However, it may be worth while for other functions to construct ones own working units since this saves time and storage.

In this chapter we will describe methods for the calculation of the function $\log(1+\frac{y}{x})$, $\arctan \frac{y}{x}$, $\sqrt{\frac{y}{x}}$; furthermore, we will deal with trigonometric, hyperbolic and exponential functions.

6.1 Calculation of logarithms

The method described below is based on the Briggs method, which was used for the first time in the calculation of logarithms to base 10 some 300 year ago:

<u>Briggs method for calculating $\log(1+\frac{y}{x})$ $(x, y > 0)$:</u>

A. Determine $q_j \in \mathbb{N}_0$ $(j = 0, 1, 2, \ldots)$ so that:

$$x + y = x \cdot \prod_{k=0}^{\infty} (1+d^{-k})^{q_k}.$$

B. Calculate $\log(1 + \frac{y}{x}) \approx \sum_{k=0}^{n-1} q_k \cdot \log(1+d^{-k})$.

Hence, the algorithm is divided into two parts. The first part (calculation of q_j) is very similar to calculation of the quotient bits in serial division and is therefore called pseudo division. The second part, calculation of

$$\sum q_k \cdot \log(1+d^{-k})$$

is carried out with table values for $\log(1+d^{-k})$ and is similar to multiplication (pseudo multiplication).

6.1.1 Calculation of q_j from $q_0, \ldots, q_{j-1}$ (pseudo division)

Assume that $Q(j-1) := \prod_{k=0}^{j-1} (1+d^{-k})^{q_k}$ has already been calculated.

q_j must be determined so that:

$$x + y \approx x \cdot \prod_{k=0}^{j} (1+d^{-k})^{q_k} = x \cdot Q(j-1) \cdot (1+d^{-j})^{q_j},$$

i.e. $y - x \cdot [Q(j-1) \cdot (1+d^{-j})^{q_j} - 1] \approx 0$.

There are several options in choosing q_j, as with division (see 4.2, 5.5); we will restrict ourselves here to the Meggitt method [Me1], which corresponds to the restoring or non-performing method as regards division.

Hence, we define:

$$y_a^{(j)} := y - x \cdot [Q(j-1) \cdot (1+d^{-j})^a - 1]$$
$$x_a^{(j)} := x \cdot Q(j-1) \cdot (1+d^{-j})^a \qquad (a = 0,1,2,\ldots)$$

and write:

$$q_j := \max \{a \mid y_a^{(j)} \geq 0\}.$$

Hence, q_j is the uniquely defined maximum value a for which the partial remainder $y_a^{(j)}$ maintains its sign.

Remark. Negative "quotient bits" q_j are also allowed in [Sa2]; as regards this method which is similar to the SRT division, the partial remainder is generally smaller than for Meggitt.

Definition $y_a^{(j)}$ and $x_a^{(j)}$ immediately yields:

Lemma 6.1. $y^{(j)}_{q_j} = y^{(j+1)}_0$; $\qquad x^{(j)}_{q_j} = x^{(j+1)}_0$;

$y^{(j)}_{a+1} = y^{(j)}_a - x^{(j)}_a \cdot d^{-j}$; $\qquad x^{(j)}_{a+1} = x^{(j)}_a + x^{(j)}_a \cdot d^{-j}$.

Proof. According to definition, we have:

$$y^{(j)}_{a+1} = y - x[Q(j-1) \cdot (1+d^{-j})^{a+1} - 1]$$

$$= y - x[Q(j-1) \cdot (1+d^{-j})^a - 1] - x \cdot Q(j-1) \cdot (1+d^{-j})^a \cdot d^{-j}$$

$$= y^{(j)}_a - x^{(j)}_a \cdot d^{-j} .$$

The remaining propositions are proved accordingly.

Hence, for determining the quotient bit q_j, we calculate

$$x^{(j)}_0, y^{(j)}_0, x^{(j)}_1, y^{(j)}_1, \ldots\ldots$$

until $y^{(j)}_i$ becomes negative. q_j is uniquely defined by:

$$y^{(j)}_{q_j} \geq 0 > y^{(j)}_{q_j+1} .$$

Restrictions of the arguments x and y

As stated, we will restrict ourselves to positive arguments. For the second part of the algorithm it is appropriate for the quotient bits to be represented with base d, i.e. $q_j < d$. This yields the following condition:

$$q_0 < d \Leftrightarrow y^{(0)}_d < 0 \Leftrightarrow y - x \cdot [1 \cdot (1+d^{-0})^d - 1] < 0 \Leftrightarrow \frac{y}{x} < 2^d - 1 .$$

It is obvious that this condition is also sufficient for $q_j < d$ for all $j \geq 0$.

6.1.2 Microprogram for the pseudo division

If
$$z_a^{(j)} := d^j \cdot y_a^{(j)},$$

we obtain the following recursion formulae (see 6.1.1):

$$\left. \begin{array}{l} z_{a+1}^{(j)} = z_a^{(j)} - x_a^{(j)} \\ x_{a+1}^{(j)} = x_a^{(j)} + x_a^{(j)} \cdot d^{-j} \end{array} \right\} \quad a = 0,1,2,\ldots,q_j-1$$

with the initial conditions:

$$z_0^{(0)} = y_0^{(0)} = y \; ; \quad x_0^{(0)} = x \; ;$$

$$z_0^{(j+1)} = d^{j+1} \cdot y_0^{(j+1)} = d^{j+1} \cdot y_{q_j}^{(j)} = d \cdot z_{q_j}^{(j)};$$

$$x_0^{(j+1)} = x_{q_j}^{(j)}.$$

$z_a^{(j)}$ can be interpreted as "partial remainder", $x_a^{(j)}$ as "divisor". The recursion formulae and the initial conditions differ only by reason of the divisor changing at every step $(x_{a+1}^{(j)} \neq x_a^{(j)})$.

The following microprogram for pseudo division is obtained by generalizing the respective program for the non-performing division (see 4.2.2, microprogram D2).

Microprogram P1 (Pseudo division)

```
0 : (DD,DE) := y ; DR := x ; Z := 0 ; q := 0 ;
1 : if DD ≥ DR
    then [DD := DD-DR; DR := DR+DR·d^-Z; q := q+1; goto 1];
    else [DE_0 := q; SHL(DD,DE); if Z < n-1 then (q := 0; Z := Z+1; goto 1)];
2 : END.
```

Remark. The shift to the left in function 1 results from the relationship

$$z_0^{(j+1)} = d \cdot z_{q_j}^{(j)} ,$$

i.e., the first partial remainder for determining q_{j+1} is obtained by a shift to the left (multiplication by d) from the last partial remainder, when calculating q_j.

6.1.3 <u>Calculation of $\sum_{k=0}^{n-1} q_k \cdot \log(1+d^{-k})$</u> (pseudo multiplication)

As with multiplication, addition of n addends is to be carried out here; we can use the algorithm for serial multiplication for this (microprogram M1, see 3.2.1), with the table values for $\log(1+d^{-k})$ taking on the role of multiplicand and the bits q_j acting as multiplier bits. Note that the "multiplicand" (contrary to serial multiplication) changes after each multiplication cycle.

<u>Microprogram P2</u> (Pseudo multiplication)

0 : Z := n-1; MP := 0; MQ = [$MQ_{n-1},...,MQ_0$] = [$q_0,...,q_{n-1}$];

1 : MD := $d^Z \cdot \log(1+d^{-Z})$; <table value>

2 : MP := MP + MD $\cdot$ MQ_0;

3 : <u>if</u> Z > 0 <u>then</u> [Z := Z-1; SHR(MP,MQ); <u>goto</u> 1];

4 : END.

Remark. Multiplication by d^Z in function 1 is necessary since Z shifts to the right (function 3) of the "partial product" MP are subsequently undertaken.

The registers MP and MD must be extended to the left by 2 sign positions to n+2 positions altogether (see Meggitt [Me 1]). The value of $\log(1+d^{-Z})$ need be stated only up to Z positions after the point. For large Z values, to sufficient accuracy:

$$\log(1+d^{-Z}) = d^{-Z}, \text{ i.e. } d^Z \cdot \log(1+d^{-Z}) = 1 .$$

Using this relationship, we can reduce the number of table values to be stored.

6.2 Calculation of arc tan $(\frac{y}{x})$

For determining arc tan $(\frac{y}{x})$ (x,y > 0), we can use a variant of the method for calculating logarithms.

Lemma 6.2. *If $q_j \in \mathbb{Z}$ are selected so that*

$$(x+iy) \cdot \prod_{k=0}^{\infty} (1-i \cdot d^{-k})^{q_k} = u \quad (u \in R, \ i := \sqrt{-1}) ,$$

then:

$$\text{arc tan} (\tfrac{y}{x}) = \text{Im } [log(x+iy)] = \sum_{k=0}^{\infty} q_k \cdot \text{arc tan}(d^{-k}).$$

Calculation of arc tan $(\frac{y}{x})$ can be carried out on the basis of this lemma as in 6.1 by pseudo division (determination of suitable q_k) and subsequent pseudo multiplication [with table values for arc tan(d^{-k})].

Proof. Let $(x+iy) \cdot \prod_{k=0}^{\infty} (1-i \cdot d^{-k})^{q_k} = u$. Then:

$$\log(x+iy) = \log u - \sum_{k=0}^{\infty} q_k \cdot \log(1-i \cdot d^{-k}) .$$

By transition to polar co-ordinates, we obtain:

$$x + iy = r \cdot e^{i \cdot \varphi} , \quad \text{i.e.} \quad \log(x+iy) = \log r + i \cdot \varphi ,$$

where $\tan \varphi = \tfrac{y}{x}$, i.e. $\varphi = \text{arc tan}(\tfrac{y}{x})$

$\Rightarrow \text{Im}[\log(x+iy)] = \varphi = \text{arc tan}(\tfrac{y}{x}).$

On the other hand:

$$\text{Im}[\log u - \sum_{k=0}^{\infty} q_k \cdot \log(1-i \cdot d^{-k})] = - \sum_{k=0}^{\infty} q_k \cdot \text{Im}[\log(1-i \cdot d^{-k})]$$

$$= - \sum_{k=0}^{\infty} q_k \cdot \text{arc tan}(\tfrac{-d^{-k}}{1}) = \sum_{k=0}^{\infty} q_k \cdot \text{arc tan}(d^{-k}).$$

Assuming that $q_0, \ldots, q_{j-1}$ have already been calculated, the next quotient bit q_j can be determined in the following manner:

We have $x_a^{(j)} + i \cdot y_a^{(j)} := (x+iy) \cdot R(j-1) \cdot (1-i \cdot d^{-j})^a$,

where $R(j-1) := \prod_{k=0}^{j-1} (1-i \cdot d^{-k})^{q_k}$

as well as $q_j := \max \{a \mid y_a^{(j)} \geq 0\}$.

Simple calculation shows that:

$$y_{a+1}^{(j)} = y_a^{(j)} - x_a^{(j)} \cdot d^{-j} \; ; \qquad y_0^{(j+1)} = y_{q_j}^{(j)} \; ; \qquad y_0^{(0)} = y \; ;$$

$$x_{a+1}^{(j)} = x_a^{(j)} + y_a^{(j)} \cdot d^{-j} \; ; \qquad x_0^{(j+1)} = x_{q_j}^{(j)} \; ; \qquad x_0^{(0)} = x \; .$$

With $z_a^{(j)} := d^j \cdot y_a^{(j)}$, this yields:

$$z_{a+1}^{(j)} = z_a^{(j)} - x_a^{(j)} \; ; \qquad z_0^{(0)} = y$$

$$x_{a+1}^{(j)} = x_a^{(j)} + z_a^{(j)} \cdot d^{-2j} \; . \qquad z_0^{(j+1)} = d \cdot z_{q_j}^{(j)}$$

As regards the respective formulae from 6.1.2, the recursion formulae and the initial conditions differ only slightly. Microprogram P1 can readily be extended to use it for calculation of arc-tangent values also. The second section of the method (pseudo multiplication) can also be performed as described in 6.1.3.

6.3 Calculation of $\sqrt{\frac{y}{x}}$ (x,y > 0)

The algorithm we are dealing with now differs substantially from the method discussed in 4.5.5 for calculation of square roots. Again, the algorithm is split into two parts:

A. q_j is determined so that:

$$\sqrt{\frac{y}{x}} = \sum_{k=0}^{\infty} q_k \cdot d^{-k} \qquad \text{(pseudo division)}.$$

B. Calculation of

$$\sum_{k=0}^{n-1} q_k \cdot d^{-k} \qquad \text{(pseudo multiplication)}.$$

Assuming that $q_0, \ldots, q_{j-1}$ are already known, we can find q_j in the following way:

We define:

$$y_a^{(j)} := y - x \cdot [S(j-1) + a \cdot d^{-j}]^2$$

$$x_a^{(j)} := 2x \cdot [S(j-1) + a \cdot d^{-j}] + x \cdot d^{-j}$$

$$z_a^{(j)} := d^j \cdot y_a^{(j)},$$

where $S(j-1) := \sum_{k=0}^{j-1} q_k \cdot d^{-k}$

and choose q_j so that:

$$q_j := \max \{a | y_a^{(j)} \geq 0\}.$$

Hence, the new partial remainder becomes as small as possible (maintaining the sign).

Simple calculation demonstrates the following recursion formulae and initial conditions:

$$\begin{array}{l|l|l}
z_{a+1}^{(j)} = z_a^{(j)} - x_a^{(j)} & z_0^{(j+1)} = d \cdot z_{q_j}^{(j)} & z_0^{(0)} = y \\
x_{a+1}^{(j)} = x_a^{(j)} + 2x \cdot d^{-j} & x_0^{(j+1)} = x_{q_j}^{(j)} - x \cdot d^{-j} + x \cdot d^{-j-1} & x_0^{(0)} = x. \\
& = x_{q_j}^{(j)} - (d-1) d^{-j-1} \cdot x &
\end{array}$$

Contrary to the methods discussed in 6.1 and 6.2, an additional correction of the "divisor" at the beginning of the calculation of q_{j+1} is necessary by subtraction of $(d-1) \cdot d^{-j-1} \cdot x$. In the binary case, this operation can be performed very quickly. The remaining steps of the algorithm correspond to those of the methods previously discussed.

6.4 Inverse functions

The following special functions can be treated by inversion of the algorithms dealt with in 6.1 - 6.3:

6.4.1 Calculation of tan(p) $[0 \le p \le \frac{\pi}{2}]$

The method consists of three parts:

A. Determination of $q_j \in \mathbb{N}_0$ with

$$p = \sum_{k=0}^{\infty} q_k \cdot \text{Arc tan}(d^{-k}) \approx \sum_{k=0}^{n-1} q_k \cdot \text{Arc tan}(d^{-k})$$

by pseudo division.

B. Investigation of x and y with

$$x + iy = u \cdot \prod_{k=0}^{n-1} (1+i \cdot d^{-k})^{q_k} \qquad (u \in \mathbb{R})$$

by pseudo multiplication.

C. Calculation of $\tan(p) \approx \frac{y}{x}$ since:

$$\text{Arc tan}(\frac{y}{x}) = \text{Im}[\log(x+iy)] = \sum_{k=0}^{\infty} q_k \cdot \text{Im}\left[\log(1+i \cdot d^{-k})\right]$$

$$\approx \sum_{k=0}^{n-1} q_k \cdot \text{Arc tan}(d^{-k}) = p \ ; \quad \text{i.e.} \quad \tan(p) \approx \frac{y}{x} \ .$$

Hence, the time necessary for tan(p) is greater than for the arc-tangent function, since additional division is required; on the contrary, when calculating arc $\tan(\frac{y}{x})$, division of the arguments at the beginning of the method can be dispensed with.

For A: we define:
$$y_a^{(j)} := p - \sum_{k=0}^{j-1} q_k \cdot \text{Arc tan}(d^{-k}) - a \cdot \text{Arc tan}(d^{-j})$$

and choose q_j similar to the previous methods so that:
$$q_j = \max \{a | y_a^{(j)} \geq 0\} ;$$

for B: we will start "multiplication" as usual with the least significant position q_{n-1} of the "multiplier".

$$x_a^{(j)} + i \cdot y_a^{(j)} := u \cdot \prod_{k=j+1}^{n-1} (1+i \cdot d^{-k})^{q_k} \cdot (1+i \cdot d^{-j})^a ;$$

$$z_a^{(j)} := d^j \cdot y_a^{(j)} \qquad (u > 0 \text{ at will}).$$

Simple calculation demonstrates the following recursion relationships:

$$z_{a+1}^{(j)} = z_a^{(j)} + x_a^{(j)} \quad ; \quad z_0^{(j-1)} = d^{-1} \cdot z_{q_j}^{(j)} ; \quad z_0^{(n-1)} = 0 ;$$

$$x_{a+1}^{(j)} = x_a^{(j)} - d^{-2j} \cdot z_a^{(j)} \quad ; \quad x_0^{(j-1)} = x_{q_j}^{(j)} \quad ; \quad x_0^{(n-1)} = u .$$

Compared with the formulae of 6.2, additions are replaced by subtractions and the shift to the right (multiplication of the partial remainder $z_{q_j}^{(j)}$ by d) by a shift to the left (division by d). Hence, the pseudo division of 6.2 becomes a pseudo multiplication.

6.4.2 Calculation of $x \cdot e^p$ or $x \cdot (e^p-1)$ [$x, p > 0$]

A. Determination of $q_j \in \mathbb{N}_0$ so that:

$$p = \sum_{k=0}^{\infty} q_k \cdot \log(1+d^{-k}) \approx \sum_{k=0}^{n-1} q_k \cdot \log(1+d^{-k})$$

using the table values for $\log(1+d^{-k})$.

B. Calculation of:

$$x \cdot (e^p - 1) \approx x \cdot [\prod_{k=0}^{n-1} (1+d^{-k})^{q_k} - 1]$$

or

$$x \cdot e^p \approx x \cdot \prod_{k=0}^{n-1} (1+d^{-k})^{q_k} \quad .$$

Writing

$$\beta := \begin{cases} 0 \text{ if } x \cdot e^p \text{ is calculated} \\ 1 \text{ if } x \cdot (e^p-1) \text{ is calculated} \end{cases}$$

and

$$y_a^{(j)} := x \cdot [\prod_{k=j+1}^{n-1} (1+d^{-k})^{q_k} \cdot (1+d^{-j})^a - \beta]$$

$$x_a^{(j)} := x \cdot [\prod_{k=j+1}^{n-1} (1+d^{-k})^{q_k} \cdot (1+d^{-j})^a]$$

$$z_a^{(j)} := d^j \cdot y_a^{(j)} \quad ,$$

we obtain the following recursion formulae and initial conditions:

$$z_{a+1}^{(j)} = z_a^{(j)} + x_a^{(j)} \quad ; \quad z_0^{(j-1)} = d^{-1} \cdot z_{q_j}^{(j)} \quad ; \quad z_0^{(n-1)} = x \cdot (1-\beta)$$

$$x_{a+1}^{(j)} = x_a^{(j)} + d^{-j} \cdot x_a^{(j)} \quad ; \quad x_0^{(j-1)} = x_{q_j}^{(j)} \quad ; \quad x_0^{(n-1)} = x \quad .$$

Comparison with the corresponding formulae from 6.4.1 demonstrates that calculation of $x \cdot e^p$ or $x \cdot (e^p-1)$ can be performed also by pseudo division and subsequent pseudo multiplication.

6.5 CORDIC method for calculating arithmetic functions

6.5.1 General description of the method

Volder [Vo1] has proposed a method for calculating trigonometric functions by a series of co-ordinate transformations. Walther [Wa2] extended this method to a more comprehensive class of functions.

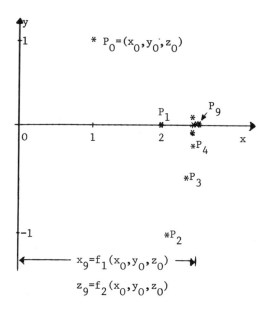

Fig. 6.1 (CORDIC transformation T_y)

The principle of this method is to transform a point (x_0, y_0, z_0) of the three-dimensional space according to a transformation rule T depending on the function to be calculated until one of the arguments of the point becomes zero, within the accuracy of calculation (see fig. 6.1; for clarity the third component is not shown there). Function values can then be obtained from the other two arguments depending on the rule of transformation chosen.

The CORDIC method (Co-ordinate Rotating Digital Computer) uses two different rules of transformation T_y and T_z, by means of which the second/third argument of the point becomes zero in n or in m steps; it is seen among other things, that trigonometric and hyperbolic functions as well as square roots can be calculated in this way. The iterations and arising functional values occurring can be described as follows:

Method T_y ($y_n \to 0$):

$$(x_0,y_0,z_0) \xrightarrow{T_y} (x_1, y_1, z_1) \xrightarrow{T_y} \ldots \xrightarrow{T_y} (x_{n-1},y_{n-1},z_{n-1}) \xrightarrow{T_y} (x_n,0,z_n)$$

$\Rightarrow x_n = f_1(x_0,y_0,z_0); \quad z_n = f_2(x_0,y_0,z_0)$.

Method T_z ($z_m \to 0$):

$$(x_0,y_0,z_0) \xrightarrow{T_z} (x'_1,y'_1,z'_1) \xrightarrow{T_z} \ldots \xrightarrow{T_z} (x'_m,y'_m,0) \quad ;$$

$\Rightarrow x'_m = f_3(x_0,y_0,z_0); \quad y'_m = f_4(x_0,y_0,z_0)$.

Functions $f_1,\ldots,f_4$ depend on the rules of transformation T_y and T_z.

6.5.2 Calculation of trigonometric functions

The rule of transformation (defined for $i \geq 0$) reads as follows:

$$\begin{cases} x_{i+1} := x_i + y_i \cdot \delta_i \\ y_{i+1} := y_i - x_i \cdot \delta_i \\ z_{i+1} := z_i + \alpha_i \end{cases} \quad \text{where} \quad \delta_i \in \mathbb{R}, \quad \alpha_i := \arctan(\delta_i)$$

[δ_i and α_i are chosen so that the argument y_n or z_n tends towards zero].

The rules of iteration stated represent a system of difference equations with the solution:

$$\begin{cases} x_{n+1} = K \cdot [x_0 \cdot \cos\alpha + y \cdot \sin\alpha] \\ y_{n+1} = K \cdot [y_0 \cdot \cos\alpha - x \cdot \sin\alpha] \\ z_{n+1} = z_0 + \alpha \end{cases} \quad \text{where} \quad \alpha := \sum_{i=0}^{n} \alpha_i \; ;$$

$$K := \prod_{i=0}^{n} \sqrt{1+\delta_i^2}.$$

One of the arguments y_n or z_n can be made to converge towards zero (for $n \to \infty$) by suitable choice of δ_i and of α_i (the two values are mutually dependent since $\alpha_i = \text{arc tan}(\delta_i)$).

For practical application of the method, iterations should be readily performed; this refers above all to multiplication by δ_i for calculating x_{i+1} or y_{i+1} from x_i and y_i. Hence, only values of the form

$$\delta_i = \pm d^{-F_i} \quad (F_i \in Z; \; d = \text{base of the notation})$$

are suitable. In this case, calculation of x_{i+1} or y_{i+1} requires only one shift over F_i digits and one addition. The necessary values $\alpha_i = \text{arc tan}(\delta_i)$ are stored in a small table.

The sign of δ_i (or of α_i) depends on the sign of x_i and y_i (or of z_i) and is chosen so that the change of y_{i+1} (or z_{i+1}) occurs towards the (x,z)- or the (x,y)-level.

For $d = 2$ we obtain the following two iteration methods:

Method T_y ($y_n \to 0$)

$$\delta_i := \begin{cases} +2^{-i} & \text{where } \text{sign}(x_i) = \text{sign}(y_i) \\ -2^{-i} & \text{where } \text{sign}(x_i) \neq \text{sign}(y_i) \end{cases}$$

$\alpha_i := \text{arc tan}(\delta_i)$.

Method T_z ($z_n \to 0$)

$$\alpha_i := \begin{cases} -\text{arc tan}(2^{-i}) & \text{where } z_i \geq 0 \\ +\text{arc tan}(2^{-i}) & \text{where } z_i < 0 \end{cases}$$

$\delta_i := \tan(\alpha_i)$.

The following lemma demonstrates that the sequence of y_i (or the sequence of z_i) tends towards zero, if the initial arguments x_0 and y_0 (or x_0 and z_0) fulfill certain conditions.

Proposition 6.3.

Let
$$\lambda_i = \begin{cases} \arctan(y_i/x_i) & \text{where } x \geq 0 \\ \pi + \arctan(y_i/x_i) & \text{where } x < 0, y \geq 0 \\ -\pi + \arctan(y_i/x_i) & \text{where } x < 0, y < 0 \end{cases}$$

i.e. $-\pi < \lambda_i \leq +\pi$; then:

a. $\lambda_{i+1} = \lambda_i - \alpha_i$; $|\lambda_{i+1}| = ||\lambda_i| - |\alpha_i||$; $|\lambda_0| \leq \sum_{j=0}^{n-1} |\alpha_j| + |\lambda_n|$

$z_{i+1} = z_i + \alpha_i$; $|z_{i+1}| = ||z_i| - |\alpha_i||$; $|z_0| \leq \sum_{j=0}^{n-1} |\alpha_j| + |z_n|$

b. If $|\alpha_i| \leq \sum_{j=i+1}^{n-1} |\alpha_j| + |\alpha_{n-1}|$, we obtain the following necessary and sufficient conditions of convergence:

$|\lambda_n| \leq |\alpha_{n-1}| \Leftrightarrow |\lambda_0| \leq \sum_{j=0}^{n-1} |\alpha_j| + |\alpha_{n-1}|$

$|z_n| \leq |\alpha_{n-1}| \Leftrightarrow |z_0| \leq \sum_{j=0}^{n-1} |\alpha_j| + |\alpha_{n-1}|$

c. The condition

$|\alpha_i| \leq \sum_{j=i+1}^{n-1} |\alpha_j| + |\alpha_{n-1}|$

is fulfilled in the binary case $[|\alpha_i| = \arctan(2^{-i})]$. For $d = 2$, we obtain:

$|\lambda_n| \leq |\alpha_{n-1}| \Leftrightarrow |\lambda_0| \leq \sum_{j=0}^{n-1} |\alpha_j| + |\alpha_{n-1}| \approx 1.74$

(for large n).

This relationship does not apply to base $d > 2$ $[|\alpha_i| = \arctan(d^{-i})]$.

Hence, the CORDIC method can be used for a non-binary base only with restriction of the range and speed of convergence.

Proof. Proposition a. can be readily proved (note that λ_i, z_i and α_i can also have negative values and that the change of values y_i and z_i occurs towards the (x,z)- or (x,y)-level).

The direction '$\Rightarrow$' of b. can be obtained directly from proposition a. For proving the inversion we will demonstrate by induction that:

$$|\lambda_i| \leq \sum_{j=i}^{n-1} |\alpha_j| + |\alpha_{n-1}| \quad (i = 0,1,\ldots) .$$

It is assumed that the initial induction holds $(i = 0)$.

If $|\lambda_i| \leq \sum_{j=i}^{n-1} |\alpha_j| + |\alpha_{n-1}|$, then

$$|\lambda_i| - |\alpha_i| \leq \sum_{j=i+1}^{n-1} |\alpha_j| + |\alpha_{n-1}| .$$

On the other hand, the following is obtained on account of the inequality assumed at the beginning of proposition b.:

$$-[|\lambda_i| - |\alpha_i|] \leq |\alpha_i| \leq \sum_{j=i+1}^{n-1} |\alpha_j| + |\alpha_{n-1}| .$$

Hence, finally:

$$|\lambda_{i+1}| = ||\lambda_i| - |\alpha_i|| \leq \sum_{j=i+1}^{n-1} |\alpha_j| + |\alpha_{n-1}| .$$

The proposition

$$|\dot{\lambda}_i| \le \sum_{j=i}^{n-1} |\alpha_j| + |\alpha_{n-1}| \qquad (i=0,1,\ldots)$$

is thus proved.

In the special case $i=n$, we have $|\lambda_n| \le |\alpha_{n-1}|$ q.e.d. For z_n, the procedure for the proof is the same (replace λ by z throughout).

Disregarding rounding errors, we obtain:

$$|\lambda_n| \quad (\text{or } |z_n|) \le \arctan(2^{-n+1}) < 2^{-n+1} \xrightarrow[n \to \infty]{} 0, \text{ if the starting}$$

values λ_0 and z_0 are within the required range.

Hence, it is sufficient to perform n+1 iterations for numbers of length n. We eliminate the influence of rounding errors by increasing the internal register length to $n + \lceil \log_2(n+1) \rceil$, since at most $\lceil \log_2(n+1) \rceil$ register positions can be falsified by rounding of intermediate results through n+1 iterations. In general, however, a much smaller number of positions are changed by rounding errors, since the roundings compensate each other at least partially.

Among other things, the most important trigonometric functions can be calculated by the two methods stated:

<u>Lemma 6.4.</u> *If the necessary conditions of convergence for the CORDIC method are met (see proposition 6.3), after n+1 iterations we obtain:*

Method T_y: $y_n \to 0$ Method T_z: $z_n \to 0$

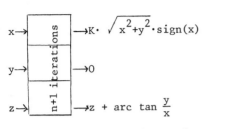

where $K := \prod_{i=0}^{n} \sqrt{1+\delta_i^2} = \prod_{i=0}^{n} \sqrt{1+2^{-2i}} \approx 1.64676$ for $n \geq 15$.

The K value depends only on the number, not on the direction of iterations; it vanishes at the output side, when multiplying the input arguments x and y by $\frac{1}{K}$.

Proof. 1. Method T_y:

According to construction we have $\lambda_{n+1} \approx 0$ also within the accuracy of calculation $y_{n+1} \approx 0$. Applying this relationship to the general solution of the difference equation system, we obtain:

$$y_{n+1} = K \cdot [y \cdot \cos \alpha - x \cdot \sin \alpha] \approx 0, \text{ i.e. } \tan \alpha \approx \frac{y}{x}.$$

$$\Rightarrow \begin{cases} z_{n+1} = z + \alpha \approx z + \arctan(\frac{y}{x}) \\ x_{n+1} \approx K \cdot \text{sign}(x) \cdot [x \cdot \frac{x}{\sqrt{x^2+y^2}} + y \cdot \frac{y}{\sqrt{x^2+y^2}}] \\ \quad\quad = K \cdot \text{sign}(x) \cdot \sqrt{x^2 + y^2} . \end{cases}$$

2. Method T_z:

$$z_{n+1} = z+\alpha \approx 0 \Rightarrow z \approx -\alpha$$

$$\Rightarrow \begin{cases} x_{n+1} \approx K \cdot [x \cdot \cos z - y \cdot \sin z] \\ y_{n+1} \approx K \cdot [y \cdot \cos z + x \cdot \sin z] \end{cases}.$$

Remark. 1. Arguments which do not satisfy the conditions of convergence must be transformed to the convergence range before starting iteration, for example by applying the relationship $\sin(2n\pi+\alpha) = \sin \alpha$.

2. Calculation of other trigonometric functions is simple: for example, we can calculate $\tan(z)$ by writing $x:=1$ and $y:=0$; we then have $x_{n+1} = K \cdot \cos z$; $y_{n+1} = K \cdot \sin z$, and $\tan(z)$ is obtained by division.

<u>Microprogram CORDIC</u> (Calculation of trigonometric functions)

<u>Remark.</u> The methods T_y and T_z differ only in the enquiry in function 1.

0 : $x :=$ argument 1; $\quad y :=$ argument 2; $\quad z :=$ argument 3; $\quad i := 0$;

1 : <u>if</u> $x \geq 0$ and $y \geq 0$ or $x \leq 0$ and $y \leq 0$ <method T_y>
     $\qquad z < 0$ <method T_z>

   <u>then</u> $[x:=x+y \cdot 2^{-i};\quad y:=y-x \cdot 2^{-i};\quad z:=z + \arctan(2^{-i})]$

   <u>else</u> $[x:=x-y \cdot 2^{-i};\quad y:=y+x \cdot 2^{-i};\quad z:=z - \arctan(2^{-i})]$;

2 : <u>if</u> $i < n$ <u>then</u> $(i:=i+1;\quad$ <u>goto</u> 1);

3 : END.

6.5.3 Other rules of transformation

Walther [Wa2] has investigated a more general class of rules of transformation by which further functions can be calculated:

$$\begin{cases} x_{i+1} := x_i + m \cdot y_i \cdot \delta_i \\ y_{i+1} := y_i - x_i \cdot \delta_i \\ z_{i+1} := z_i + \alpha_i \end{cases} \quad \text{where} \quad m \in \{-1, 0, 1\}, \quad \delta_i \in \mathbb{R},$$

$$\alpha_i := \begin{cases} \arctan(\delta_i) & m = 1 \\ \delta_i & m = 0 \\ \text{arc tanh}(\delta_i) & m = -1 \end{cases}$$

Solution of this system of difference equation reads:

$$\begin{cases} x_{n+1} = K_m \cdot [x_0 \cdot \cos(\alpha \cdot \sqrt{m}) + y_0 \cdot \sqrt{m} \cdot \sin(\alpha \cdot \sqrt{m})] \\ y_{n+1} = K_m \cdot [y_0 \cdot \cos(\alpha \cdot \sqrt{m}) - x_0 \cdot \frac{1}{\sqrt{m}} \sin(\alpha \cdot \sqrt{m})] \\ z_{n+1} = z_0 + \alpha, \end{cases}$$

where $\alpha := \sum_{i=0}^{n} \alpha_i$,

$$K_{+1} := \prod_{i=0}^{n} \sqrt{1+\delta_i^2}, \quad K_0 := 1, \quad K_{-1} := \prod_{i=0}^{n} \sqrt{1-\delta_i^2}.$$

The rules for $m = -1, 0, +1$ require in each case different parameters δ_i, independent of m. As in 6.5.2, these values must be chosen so that one of the arguments y_{n+1} or z_{n+1} becomes zero after $n+1$ iterations within the accuracy of calculation. Details regarding this can be found in [Wa2].

If the y_i or z_i sequence converges towards zero, the following functional values can be read off at the two remaining outputs:

I. $\underline{m = +1}$ (trigonometric functions and roots, see 6.5.2)

II. $\underline{m = 0}$ (multiplication and division)

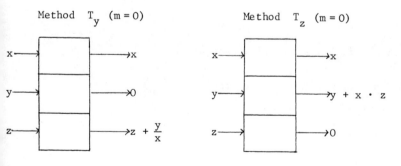

III. $m = -1$ (hyperbolic functions and roots)

Method T_y

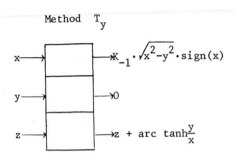

Method T_z

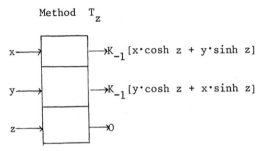

It is evident that, apart from calculation of special functions, the CORDIC technique can be used also for elementary arithmetic operations. Further functions can be simply calculated from the results provided by CORDIC, for example:

$$\tanh(u) = \frac{\sinh(u)}{\cosh(u)} \quad ;$$

$$e^u = \frac{e^u + e^{-u}}{2} + \frac{e^u - e^{-u}}{2} = \sinh(u) + \cosh(u) \quad ;$$

$$\log u = 2 \cdot \text{arc tanh} \left(\frac{u-1}{u+1}\right) \quad [\text{since } \tanh(u) = \frac{e^{2u} - 1}{e^{2u} + 1}] \quad ;$$

$$\sqrt{u} = \sqrt{(u + \tfrac{1}{4})^2 - (u - \tfrac{1}{4})^2} \quad .$$

6.5.4 Examples of the CORDIC method

The CORDIC method for $m = +1, 0, -1$ is very efficient, when parallel calculation is possible of $(x_{i+1}, y_{i+1}, z_{i+1})$ from (x_i, y_i, z_i). The mircoprogram CORDIC, explained in 6.5.2, functions in the form stated for $m = +1$ only. In order to allow application for other values of m, the variable i is to be replaced by F_i throughout in function 1, while F_i can be chosen as follows:

a. $m = +1$: $F_i = i$

b. $m = 0$: $F_i = i-1$

c. $m = -1$: $(F_0, F_1, F_2, \ldots)$
$$= (1,2,3,4,4,5,6,7,8,9,10,11,12,13,13,14,15,\ldots).$$

In this case, the F_i sequence consists of natural numbers, while the numbers
$4, 13, 40, 121, \ldots, k, 3k+1, \ldots$ occur twice.

6.5.4.1 CORDIC transformation for $m = +1$

For $n \geq 15$ (n+1 = number of iteration steps), we have:
$$K_{+1} \approx 1.64676$$

This K_{+1} value will be used below.

A. Method T_y

We are setting: $x_0 = y_0 = 1$, $z_0 = 0$. Thus:

$x_i \longrightarrow K_{+1} \cdot \sqrt{x_0^2 + y_0^2} \cdot \text{sign } x_0 = K_{+1} \cdot \sqrt{2} \approx 2.3288707$.

$z_i \longrightarrow z_0 + \arctan(\frac{y_0}{x_0}) = \arctan(1) = \frac{\pi}{4} \approx 0.78539816$.

Table 6.1

i	x_i	y_i	z_i
0	1.000000	1.000000	0.000000
1	2.000000	0.000000	0.785398
2	2.000000	-1.000000	1.249046
3	2.225000	-0.500000	1.004067
4	2.312500	-0.218750	0.879122
5	2.326172	-0.074219	0.817293
6	2.328491	-0.001526	0.786053
7	2.328515	0.034857	0.770430
8	2.328787	0.016665	0.778242
9	2.328852	0.007568	0.782148
10	2.328867	0.003020	0.784101
11	2.328870	0.000746	0.785078
12	2.328871	-0.000392	0.785566
13	2.328871	0.000177	0.785322
14	2.328871	-0.000107	0.785444
15	2.328871	0.000035	0.785383
16	2.328871	-0.000036	0.785414
17	2.328871	-0.000001	0.785398
18	2.328871	0.000017	0.785391
19	2.328871	0.000008	0.785395
20	2.328871	0.000004	0.785397
21	2.328871	0.000002	0.785397
22	2.328871	0.000000	0.785398

B. <u>Method T_z</u>

Here we are setting: $x_0 = -1$, $y_0 = 1$, $z_0 = 0$.

$x_i \longrightarrow K_{+1} \cdot [x_0 \cdot \cos z_0 - y_0 \cdot \sin z_0] = K_{+1} \cdot x_0 \approx -1.646760258$.

$y_i \longrightarrow = K_{+1} \cdot y_0 \approx +1.646760258$.

Table 6.2

i	x_i	y_i	z_i
0	-1.000000	1.000000	0.000000
1	-2.000000	0.000000	-0.785398
2	-2.000000	1.000000	-0.321751
3	-1.750000	1.500000	-0.076772
4	-1.562500	1.718750	0.047583
5	-1.669122	1.621094	-0.014836

i	x_i	y_i	z_i	(continuation)
6	-1.619263	1.673279	0.016404	
7	-1.645408	1.647978	0.000780	
8	-1.658283	1.635123	-0.007032	
9	-1.651895	1.641601	-0.003126	
10	-1.648689	1.644827	-0.001173	
11	-1.647083	1.646437	-0.000196	
12	-1.646279	1.647241	0.000292	
13	-1.646681	1.646839	0.000048	
14	-1.646882	1.646638	-0.000074	
15	-1.646782	1.646739	-0.000012	
16	-1.646731	1.646789	0.000018	
17	-1.646756	1.646764	0.000002	
18	-1.646769	1.646752	0.000005	
19	-1.646763	1.646758	-0.000001	
20	-1.646760	1.646761	-0.000000	
21	-1.646761	1.646759	0.000000	
22	-1.646760	1.646760	-0.000000	

6.5.4.2 CORDIC transformation for $m = -1$

For $n \geq 15$ (n+1 = number of iteration steps), we have:

$$K_{-1} \approx 0.82816 \,.$$

A. Method T_y

($x_0 = 1.5$, $y_0 = -1$, $z_0 = 0$)

$$x_i \longrightarrow K_{-1} \cdot \sqrt{x_0^2 - y_0^2} \cdot \text{sign } x_0 = K_{-1} \cdot \sqrt{1.25} \approx 0.9259103 \,.$$

$$z_i \longrightarrow z_0 + \text{arc tanh}\left(\frac{y_0}{x_0}\right) = \text{arc tanh}(-1.5) \approx -0.80471895 \,.$$

Table 6.3

i	x_i	y_i	z_i
0	1.500000	-1.000000	0.000000
1	1.000000	-0.250000	-0.549306
2	0.937500	0.000000	-0.804719
3	0.937500	-0.117188	-0.679062
4	0.930176	-0.058594	-0.741643
5	0.926514	-0.000458	-0.804225

i	x_i	y_i	z_i	(continuation)
6	0.926500	0.028496	−0.835485	
7	0.926054	0.014019	−0.819859	
8	0.925945	0.006784	−0.812046	
9	0.925918	0.003167	−0.808140	
10	0.925912	0.001359	−0.806187	
11	0.925911	0.000458	−0.805210	
12	0.925910	0.000003	−0.804722	
13	0.925910	−0.000223	−0.804478	
14	0.925910	−0.000110	−0.804600	
15	0.925910	0.000003	−0.804722	
16	0.925910	−0.000054	−0.804661	
17	0.925910	−0.000026	−0.804691	
18	0.925910	−0.000011	−0.804707	
19	0.925910	−0.000004	−0.804714	
20	0.925910	−0.000001	−0.804718	
21	0.925910	0.000001	−0.804720	
22	0.925910	0.000000	−0.804719	

B. Method T_z

($x_0 = -1$, $y_0 = -1$, $z_0 = 0$)

$x_i \longrightarrow K_{-1} \cdot [x_0 \cdot \cosh z_0 + y_0 \cdot \sinh z_0] = K_{-1} \cdot x_0 \approx -0.82815936$.

$y_i \longrightarrow K_{-1} \cdot [y_0 \cdot \cosh z_0 + x_0 \cdot \sinh z_0] = K_{-1} \cdot y_0 \approx -0.82815936$.

Table 6.4

i	x_i	y_i	z_i
0	−1.000000	−1.000000	0.000000
1	−1.500000	−1.500000	−0.549306
2	−1.112500	−1.112500	−0.293893
3	−0.984238	−0.984238	−0.168236
4	−0.922282	−0.922282	−0.105655
5	−0.865173	−0.865173	−0.043073
6	−0.838137	−0.838137	−0.011813
7	−0.825041	−0.825041	0.003813
8	−0.831487	−0.831487	−0.003999
9	−0.828238	−0.828238	−0.000093
10	−0.826621	−0.826621	0.001860
11	−0.827428	−0.827428	0.000884
12	−0.827823	−0.827823	0.000395
13	−0.828034	−0.828034	0.000151
14	−0.828135	−0.928135	0.000029
15	−0.828236	−0.828236	−0.000093

i	x_i	y_i	z_i	(continuation)
16	-0.828186	-0.828186	-0.000032	
17	-0.828160	-0.828160	-0.000001	
18	-0.828148	-0.828148	0.000014	
19	-0.828154	-0.828154	0.000006	
20	-0.828157	-0.828157	0.000002	
21	-0.828159	-0.828159	0.000001	
22	-0.828160	-0.828160	-0.000000	

6.5.4.3 <u>Graphic illustration of examples of the CORDIC method</u>

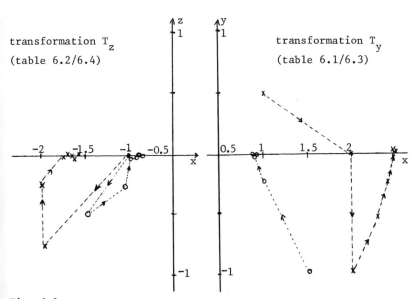

transformation T_z
(table 6.2/6.4)

transformation T_y
(table 6.1/6.3)

Fig. 6.2

o⋯→⋯o⋯⋯→⋯o⋯⋯→⋯o⋯⋯→⋯o   CORDIC method m = -1
x--→-x--→-x--→-x--→-x   CORDIC method m = +1

7. Time complexity of arithmetic operations

7.1 Description of the model

In this last chapter, upper and lower limits for the cycle time for calculation of arithmetic operations are derived. In order to compare the results, it is assumed that the functions are constructed with circuits comprising only denary circuit elements with r inputs at most (called (d,r)-elements).

It is assumed furthermore that all (d,r)-elements can be obtained by the same effort, that they require the same cycle time (one time unit or logical step) and that they are not subject to fan-out restrictions. In practice, these model conditions can be met only approximately.

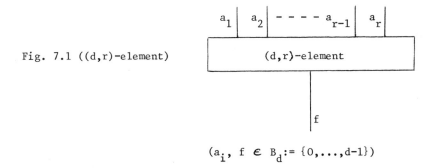

Fig. 7.1 ((d,r)-element)

$(a_i, f \in B_d := \{0,...,d-1\})$

Example. There are 16 different (2,2)-elements (binary gates with 2 inputs at most). We can, for example, regard the AND or OR gate with one or two inputs and negation as (2,2)-elements. Contrary to our previous cycle time conditions, it is assumed in this chapter that the cycle time for negation cannot be disregarded. This does not generally result in an increase in the cycle time, since negation can almost always be taken into the next or previous module, and does not therefore have to be stated as an individual circuit element.

A circuit of (d,r)-elements (i.e. interconnection of (d,r)-elements without feedback), with m inputs and k outputs, can be described by:

$$S : B_d^m \to B_d^k .$$

All inputs into (d,r)-elements which are not at the same time outputs of another (d,r)-element of S are called <u>inputs of S</u>; the <u>outputs</u> of the (d,r)-circuit are defined similarly. The <u>number of steps of a circuit output</u> is the maximum number of (d,r)-elements which have to be passed in sequence.

The <u>number of steps (cycle time) of a circuit</u> is defined as the maximum number of steps of the circuit outputs.

In order to calculate an arithmetic operation employing a (d,r)-circuit, the arguments of the operation must first be coded to denary form; the result of the operation is obtained by decoding at the outputs of the circuit.

<u>Definition 7.1.</u> *Let* $\Phi : X_1 \dot{\times} \ldots \dot{\times} X_n \to Y$ *be an n-digit (arithmetic) operation; it is assumed that* X_i *(*$i = 1,\ldots,n$*) and Y are finite sets.*

A τ-*step* <u>*denary circuit*</u> *with m inputs and k outputs (i,e, a (d,r)-circuit* $S_\tau : B_d^m \to B_d^k$ *with number of steps* τ*)* <u>*calculates (the result of the operation)*</u> Φ*, if:*

a. *There are* <u>*codings*</u> $c_i : X_i \to B_d^{u_i}$ *of the arguments* X_i *with*

$$c_1(X_1) \dot{\times} \ldots \dot{\times} c_n(X_n) \subset B_d^{u_1} \dot{\times} \ldots \dot{\times} B_d^{u_n} \subset B_d^m$$

b. *There is a* <u>*bijective decoding*</u>

$$h' : \bigcup_{\substack{(x_1,\ldots,x_n) \\ \in X_1 \dot{\times} \ldots \dot{\times} X_n}} S_\tau(c_1(x_1),\ldots,c_n(x_n)) \to Y$$

with the following characteristics:

For all $(x_1,\ldots,x_n) \in X_1 \dot{\times} \ldots \dot{\times} X_n$, *we have:*

$$S_\tau(c_1(x_1),\ldots,c_n(x_n)) = h'^{-1}(\Phi(x_1,\ldots,x_n)),$$

i.e. the circuit provides the correct — uniquely decodable — result $\Phi(x_1,\ldots,x_n)$ of the operation Φ for all permissible input combinations $(x_1,\ldots,x_n)$.

Figure 7.2 clarifies the contents of this definition (see 7.2.1 for definition of $h_k(y)$).

<u>Remarks.</u> 1. Coding and decoding times are disregarded. This is justified if the respective cycle time is small as compared with the cycle time τ of the circuit. Furthermore, this provides limits for the "pure" cycle time of the circuit.

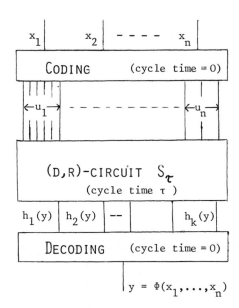

Fig. 7.2
(Calculation of Φ with S_τ)

$$(x_1,\ldots,x_n) \xrightarrow{c_i} (c_1(x_1),\ldots,c_n(x_n))$$

$$\Phi \downarrow \qquad\qquad\qquad \downarrow S_\tau$$

$$\Phi(x_1,\ldots,x_n) \underset{h'^{-1}}{\overset{h'}{\rightleftarrows}} S_\tau(c_1(x_1),\ldots,c_n(x_n))$$

S_τ calculates Φ, if the diagram is commutative.

2. The limits derived in this chapter are valid only for non-redundant notations because of the assumed bijectivity of the decodings. Completely disregarding unambiguity of representation of $\Phi(x_1,\ldots,x_n)$ (i.e. bijectivity of decoding), a trivial circuit with the number of steps $\tau = 0$ could be employed throughout simply by postponing calculation of $\Phi(x_1,\ldots,x_n)$ to the decoding (which is supposed to be done in zero time).

Reasonable lower limits for the cycle time of arithmetic operations are at present unknown for redundant notations and/or taking the decoding time into consideration.

3. The assumed bijectivity of decoding

$$h' : B_d^k \supset S_\tau(c_1(X_1),\ldots,c_n(X_n)) \to Y$$

is equivalent to the demand of the existence of a <u>mapping</u>

$$h : Y \to B_d^k ,$$

i.e. with non-redundancy of representation of all $y \in Y$. In this case, we have:

$$h(y) = h'^{-1}(y) \quad \text{for all } y \in Y .$$

7.2 Lower limits of cycle time for arithmetic operations

In order to derive lower limits for the cycle time of an (arithmetic) operation, we will examine how many inputs can influence one output. This provides information on how many (d,r)-elements with series connection are needed at a minimum to calculate the result of the operation in non-redundant denary form.

7.2.1 Separability of the circuit outputs

Definition 7.2. *Let S_τ calculate $\Phi: X_1 \dot{\times} \ldots \dot{\times} X_n \to Y$ in τ time units.*

1. *$h_i(y)$ denotes the i-th output ($i = 1,\ldots,k$) of S_τ for the argument $y = \Phi(x_1,\ldots,x_n)$, i.e.*

$$h(y) = (h_1(y),\ldots,h_k(y)) := S_\tau(c_1(x_1),\ldots,c_n(x_n)).$$

2. *$A_q^{(j)} \subset X_q$ is called $\underline{h_j\text{-separable regarding } X_q}$, if:*

For any $a_q^{(1)} \ne a_q^{(2)} \in A_q^{(j)}$ there are $x_1,\ldots,x_{q-1},x_{q+1},\ldots,x_n$ with

$$h_j(\Phi(x_1,\ldots,x_{q-1},a_q^{(1)},x_{q+1},\ldots,x_n)) \ne h_j(\Phi(x_1,\ldots,x_{q-1},a_q^{(2)},x_{q+1},\ldots,x_n)).$$

Hence, if $A_q^{(j)}$ is h_j-separable as regards X_q, h_j depends at least on $|A_q^{(j)}|$ values of the input argument X_q of Φ. Since X_q has been coded in denary numbers, h_j depends at least on $\log_d |A_q^{(j)}|$ input <u>lines</u> of S_τ; all these lines are part of the coding of $c_q(X_q)$. Since this applies to all arguments X_q ($q = 1,\ldots,n$), every j-th output of the circuit really depends at least on

$$\sum_{i=1}^{n} \lceil \log_d |A_i^{(j)}| \rceil \quad \text{inputs.}$$

On the other hand it is evident that each output really depends on r^τ inputs only, since only those circuit elements are used with r inputs at most, and since the output is calculated in τ steps at most (see fig. 7.3).

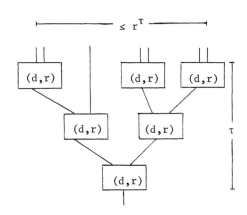

Fig. 7.3

We obtain finally:

$$r^\tau \geq \max_j \{ \sum_{i=1}^n \lceil \log_d |A_i^{(j)}| \rceil \},$$

yielding the following estimate for the cycle time τ:

Lemma 7.1. *If S_τ is a (d,r)-circuit for calculating*

$$\Phi : X_1 \dot\times \ldots \dot\times X_n \to Y \quad \text{with k output lines}$$

and if $A_i^{(j)}$ is an h_j-separable part quantity of X_i ($i = 1,\ldots,n$; $j = 1,\ldots,k$), then:

$$\tau \geq \max_{j \in \{1,\ldots,k\}} \{ \lceil \log_r (\sum_{i=1}^n \lceil \log_d |A_i^{(j)}| \rceil) \rceil \} .$$

We will demonstrate application of this central result on some examples.

7.2.2 'Examples

A. <u>Addition of two non-negative integer numbers (random coding of inputs; output in binary digit coding)</u>

The operation Φ has the following form:

$$\Phi : \underbrace{[0 : 2^n-1]}_{X_1} \dot{\times} \underbrace{[0 : 2^n-1]}_{X_2} \to Y$$

with the following interpretation ⓐ or ⓑ :

ⓐ $\Phi(a,b) := a+b$, i.e. $Y = [0 : 2^{n+1}-2]$;

ⓑ $\Phi(a,b) := \begin{cases} a+b & \text{when } a+b < 2^n \\ a+b-2^n & \text{otherwise} \end{cases} = a+b \bmod 2^n$; i.e. $Y = X_1$.

The result is to be given in denary digit coding; allocating the value 2^{i-1} to the output h_i, we obtain:

$$h(\Phi(a,b)) = h(y) = \begin{cases} (h_{n+1}(y), h_n(y), \ldots, h_1(y)) & \text{where ⓐ} \\ (h_n(y), \ldots, h_1(y)) & \text{where ⓑ} \end{cases}$$

For ⓐ we require one more binary digit than for ⓑ .

<u>Lemma 7.2.</u> 1. X_1 and X_2 are h_n-*separable*.

2. *For minimum cycle time* τ *of a* (d,r)-*circuit for calculating* Φ, *we have*:

$$\tau \geq \lceil \log_r (2 \cdot \lceil n \cdot \log_d 2 \rceil) \rceil.$$

Proof. Due to commutativity of operation Φ it is sufficient to prove proposition 1 for the set X_1. Demand 2 follows from lemma 7.1.

We have to prove that h_n (i.e. the digit of the result having the value 2^{n-1}) actually depends on all values of the first addend. We assume that $a,b \in X_1$ and $a < b$. We then have to differentiate three cases:

1. $0 \le a < b \le 2^{n-1} - 1$:

With $z = 2^{n-1} - 1 - a \in X_2$, we have due to $b > a$:

$$h_n(\Phi(a,z)) = h_n(2^{n-1}-1) = 0$$
$$\neq h_n(\Phi(b,z)) = h_n(2^{n-1}-1+(b-a)) = 1.$$

2. $2^{n-1} \le a < b \le 2^n - 1$:

For $z = 2n + (2^{n-1} - 1 - a) \in X_2$, we have:

$$h_n(\Phi(a,z)) = h_n(2^n + (2^{n-1}-1)) = 0$$
$$\neq h_n(\Phi(b,z)) = h_n(2^n + 2^{n-1} - 1 + (b-a)) = 1.$$

3. $a \le 2^{n-1} - 1;\ b \ge 2^{n-1}$:

$$\Rightarrow h_n(\Phi(a,0)) = 0 \neq h_n(\Phi(b,0)) = 1,$$

thus demonstrating h_n-separability of X_1 (and of X_2).

$$\Rightarrow |A_1^{(n)}| = |A_2^{(n)}| = |X_1| = |X_2| = 2^n,$$

i.e. $\quad \tau \ge \lceil \log_r(\lceil \log_d(2^n) \rceil + \lceil \log_d(2^n) \rceil) \rceil = \lceil \log_r(2 \cdot \lceil n \cdot \log_d 2 \rceil) \rceil.$

In the special case, d = 2, this is simplified to:

$$\tau \geq \lceil \log_r(2n) \rceil \ .$$

Remark. 1. We will demonstrate that the lower limit stated $\lceil \log_r(2n) \rceil$ for the addition of two n-digit binary numbers with (2,r)-elements applies <u>to all codings</u> of the circuit outputs.

2. The conditional sum adder (see 2.5) has the following cycle time:

$$\tau_{\text{COND SUM}} = 2 \cdot (\lceil \log_2 n \rceil + 2) \ .$$

Two of the time units (i.e. selection of the final result depending on whether addition or subtraction was carried out) are unnecessary if we restrict ourselves to addition, as shown in this example; this yields the following estimated cycle time for the duration of a conditional sum addition:

$$\tau_{\text{COND SUM}}^{\text{ADD}} = 2 \cdot (\lceil \log_2 n \rceil + 1) \ .$$

The lower limit for an adder, consisting of (2,2)-elements (the conditional sum adder uses only modules with 2 inputs at most) is as follows:

$$\tau \geq \lceil \log_2(2n) \rceil = \lceil \log_2 n \rceil + 1 \ .$$

Hence, the cycle time of the conditional sum adder is greater than the general limit by the factor 2 only, but it is seen that the adder does not at all use the majority of the (2,2)-elements (for example the modulo 2 sum of inputs).

The carry select adder (see 2.6) also has a cycle time very close to the lower limit.

B. Size comparison of two numbers

Comparison of two numbers can be described by the following operation Φ:

$$\Phi : \underbrace{[0 : N-1]}_{X_1} \times \underbrace{[0 : N-1]}_{X_2} \to B$$

$$\Phi(a,b) := \begin{cases} 0 & \text{where } a < b \\ 1 & \text{where } a \geq b. \end{cases}$$

Lemma 7.3. *A (d,r)-circuit S_τ for the calculation of Φ has a minimum cycle time of:*

$$\tau \geq \lceil \log_r 2 \lceil \log_d N \rceil \rceil \quad \text{time units.}$$

Proof. There is at least one circuit output h_j, for which:

$$h_j(0) \neq h_j(1)$$

(otherwise S_τ could not calculate function Φ).
X_1 (hence also X_2 for reasons of symmetry) is h_j-separable, since for $a,b \in X_1$ with $a < b$, we have:

$$h_j(\Phi(a,b)) = h_j(0) \neq h_j(1) = h_j(\Phi(b,b)).$$

The proposition is proved since $|X_1| = N$

C. Calculation of the integer part of a product

In this case, the function Φ can be defined as follows:

$$\Phi : [0 : N-1] \,\dot\times\, [0 : N-1] \to [0 : N-1]$$

$$\Phi(a,b) := \lfloor \tfrac{a \cdot b}{N} \rfloor \; .$$

Lemma 7.4. *A (d,r)-circuit for calculating Φ requires:*

$$\tau \geq \lceil \log_r 2 \lceil \log_d \lfloor \sqrt{N} \rfloor \rceil \rceil \quad \textit{logical steps.}$$

<u>Proof.</u> We will consider an output h_j of the circuit, for which

$$h_j(0) \neq h_j(1) \ .$$

Such an output always exists. Hence, it is sufficient to demonstrate that the part quantity

$$[1 : \lfloor \sqrt{N} \rfloor] \quad \text{of} \quad [0 : N-1] \quad \text{is } h_j\text{-separable.}$$

Assuming $0 \leq a < b \leq \lfloor \sqrt{N} \rfloor$, then there exists a value $z \in [0 : N-1]$ with $a \cdot z < N \leq b \cdot z < 2N$;

i.e. $h_j(\Phi(a,z)) = h_j(0) \neq h_j(1) = h_j(\Phi(b,z))$.

Q.e.d.

7.2.3 Lower limits

The lower limit for addition time derived in 7.2.2 depends on the coding of the circuit outputs. Winograd ([Wi2], [Wi3]) and Spira [Sp1] have derived limits which do not depend on this coding. The following is a collation of the main results of these operations (without proofs).

<u>Definition 7.3.</u> *1.* $\alpha(\mu) := \max \{p^n | p \text{ prime}, n \in \mathbb{N}, p^n \text{ divides } \mu\}$;

2. $Q_m := kgV(1,2,\ldots,m); \quad \gamma(\mu) := \min \{m | Q_m \geq \mu\}$;

3. $\beta(\mu) := \begin{cases} 2^{n-1} & \text{when } \mu = 2^n, n < 3 \\ 2^{n-2} & \text{when } \mu = 2^n, n \geq 3 \\ \max\{p^{n-1}, \alpha(p-1)\} & \text{when } \mu = p^n; p \text{ prime}, p \neq 2 \\ \max\{\beta(p^j) | p \text{ prim and } p^j \text{ divides } \mu\} & \text{otherwise} \end{cases}$

It is evident that $\gamma(\mu)$ must be a power of a prime number, i.e.

$$\gamma(\mu) = q^j \qquad (q \text{ prime}, j \in \mathbb{N}_0).$$

We will examine the following functions:

$\Phi_1 : [0:N-1] \dot\times [0:N-1] \to [0:N-1]$ with $\Phi_1(a,b) := a+b \mod N$
$\Phi_2 : [0:N-1] \dot\times [0:N-1] \to [0:2N-2]$ with $\Phi_2(a,b) := a+b$
$\Phi_3 : [0:N-1] \dot\times [0:N-1] \to [0:N-1]$ with $\Phi_3(a,b) := a \cdot b \mod N$
$\Phi_4 : [1:N] \dot\times [1:N] \to [1:N^2]$ with $\Phi_4(a,b) := a \cdot b$.

The following limits apply to the cycle time of this operation:

<u>Proposition 7.5.</u>

$\tau_{\Phi_1} \geq \lceil \log_r 2 \cdot \lceil \log_d \alpha(N) \rceil \rceil$; $\tau_{\Phi_2} \geq \lceil \log_r 2 \cdot \lceil \log_d \gamma(\lceil \frac{N}{2} \rceil) \rceil \rceil$;

$\tau_{\Phi_3} \geq \lceil \log_r 2 \cdot \lceil \log_d \beta(N) \rceil \rceil$; $\tau_{\Phi_4} \geq \lceil \log_r 2 \cdot \lceil \log_d \gamma(\lceil \frac{\lfloor \log_2 N \rfloor + 1}{2} \rceil) \rceil \rceil$.

<u>Examples.</u> 1. $N = 2^n$ ($n \geq 3$), $d = 2$ $\Rightarrow$ $\alpha(N) = 2^n$; $\beta(N) = 2^{n-2}$;

i.e. $\tau_{\Phi_1} \geq \lceil \log_r(2n) \rceil$; $\tau_{\Phi_3} \geq \lceil \log_r 2(n-2) \rceil$.

2. $N = 2^{10}$, $d = 2$ $\Rightarrow$ $\gamma(\lceil \frac{\lfloor \log_2 N \rfloor + 1}{2} \rceil) = \gamma(6) = 3$;

$\gamma(\lceil \frac{N}{2} \rceil) = \gamma(512) = 8$, since $\text{kgV}(1,..,8) = 840$; $\text{kgV}(1,..,7) = 420$;

i.e. $\tau_{\Phi_1} \geq \lceil \log_r 20 \rceil$; $\tau_{\Phi_2} \geq \lceil \log_r 6 \rceil$; $\tau_{\Phi_3} \geq \lceil \log_r 16 \rceil$; $\tau_{\Phi_4} \geq \lceil \log_r 4 \rceil$.

In the special case $r = 2$ (i.e. circuit elements with two inputs at most), this becomes:

$$\tau_{\Phi_1} \geq 5; \quad \tau_{\Phi_2} \geq 3; \quad \tau_{\Phi_3} \geq 4; \quad \tau_{\Phi_4} \geq 2 .$$

Remark. In 7.2.2 (example A.) we derived

$$\tau \geq \lceil \log_r(2n) \rceil \quad \text{as the lower limit for } N = 2^n \text{ for the two functions } \Phi_1 \text{ and } \Phi_2 .$$

We obtained this limit for a special output coding.

Proposition 7.5 shows that we can indicate a lower limit for the function Φ_2 by choosing a different coding of the outputs (we will find in 7.3 that there is a circuit with cycle time exceeding this limit only slightly); as regards the function Φ_1, improvement of the lower limit is, however, impossible.

7.3 Upper limits

Winograd has provided circuits for the operations $\Phi_1, \ldots, \Phi_4$ defined in 7.2.3, the cycle time of which approaches the given lower limit. These results were extended in a work by Spira [Sp1] based on this:

Proposition 7.6. *There are (d,r)-circuits (and respective codings of the circuit inputs or outputs) calculating the functions Φ_i in t_{Φ_i} time units, with t_{Φ_i} as follows:*

$$t_{\Phi_1} = 1 + \left\lceil \log_r \left\lceil \frac{\lceil \log_d \alpha(N) \rceil}{\lfloor \frac{r}{2} \rfloor} \right\rceil \right\rceil ;$$

$$t_{\Phi_2} = 1 + \left\lceil \log_r \left\lceil \frac{\lceil \log_d \gamma(2N-1) \rceil}{\lfloor \frac{r}{2} \rfloor} \right\rceil \right\rceil ;$$

$$t_{\Phi_3} = t_{\Phi_1}$$

$$t_{\Phi_4} = 1 + \left\lceil \log_r \left\lceil \frac{\lceil \log_d \gamma(2\lfloor \log_2 N \rfloor - 1) \rceil}{\lfloor \frac{r}{2} \rfloor} \right\rceil \right\rceil .$$

Comparing this with the results of 7.2.3, it is shown that the cycle times t_{Φ_i} are only slightly higher than the lower limits for τ_{Φ_i}.

Example. $N = 2^n$, $d = 2$;

$\Rightarrow t_{\Phi_1} = 1 + \lceil \log_r \lceil \frac{n}{\lfloor \frac{r}{2} \rfloor} \rceil \rceil \leq 1 + \lceil \log_r n \rceil = \lceil \log_r rn \rceil$.

On the other hand, we have in general:

$\tau_{\Phi_1} \geq \lceil \log_r 2n \rceil$.

The limits differ at most by one time unit.

Remarks. 1. The codings of the in/outputs for the Spira circuits for calculating Φ_i in t_{Φ_i} time units depend on the function Φ_i. No coding is known which can be applied to addition and multiplication at the same time without at least one of the two circuits having a considerably longer cycle time than the respective Spira circuit. If $N = 2^n$, the Spira circuits have at least 2^n output lines; application of these circuits is therefore impossible in practice.

2. Both the lower and upper limits for multiplication are lower than the respective values for addition; the same applies to the limits of the "modulo N" operations Φ_1 and Φ_3, when compared with the operations Φ_2 or Φ_4. These results, which might seem surprising at first, are explained by the definition of the word "calculation" (see definition 7.1):

If $\Phi : X_1 \dot{\times} X_2 \to Y$ is an arithmetic operation, only the following separation of classes is carried out by the Winograd/Spira circuits S_{τ_Φ} :

$X_1 \dot{\times} X_2 \xrightarrow{S_{\tau_\Phi}} X_1 \dot{\times} X_2 \big|_\sim$.

where $(a_1, b_1) \sim (a_2, b_2) \Leftrightarrow \Phi(a_1, b_1) = \Phi(a_2, b_2)$.

For calculation of the function Φ it is sufficient to allocate one common output line to all elements of one class (i.e. all combinations of arguments for which the operation Φ yields the same result). The Winograd and Spira circuits function on this principle. In general, such splitting into classes is easier and quicker for multiplication than for addition; note that the circuit <u>expenditure</u> is unimportant as regards <u>time complexity</u>.

7.4 Calculation of functions Φ_1 and Φ_2 (addition) with binary digit coding of the in/outputs of the circuit

Simplicity of the design principle of the Winograd and Spira circuits results in very fast, but far too costly constructions. We are very interested therefore in the question as to whether or not there are (d,r)-circuits which carry out calculation with an acceptable expenditure (as regards number of modules or outputs of the circuit) and which do not, at the same time, essentially exceed the lower limit for the cycle time.

Brent [Br1] has demonstrated that there are such circuits for binary addition (i.e. for calculation of functions Φ_1 and Φ_2):

$$\Phi_1 : [0:2^n-1] \times [0:2^n-1] \rightarrow [0:2^n-1] \quad \text{with} \quad \Phi_1(a,b) := a+b \bmod 2^n$$

$$\Phi_2 : [0:2^n-1] \times [0:2^n-1] \rightarrow [0:2^{n+1}-2] \quad \text{with} \quad \Phi_2(a,b) := a+b .$$

The operations Φ_i are to be calculated on a $(2,r)$-circuit S_{τ_n} with <u>binary digit coding</u> of the in/outputs.

We have:

$$a = \sum_{i=0}^{n-1} a_i \cdot 2^i, \quad b = \sum_{i=0}^{n-1} b_i \cdot 2^i$$

$$s = \begin{cases} \sum_{i=0}^{n-1} s_i \cdot 2^i = a+b \bmod 2^n \\ \sum_{i=0}^{n} s_i \cdot 2^i = a+b \ . \end{cases}$$

The results of 7.3 show that every circuit S_τ requires a cycle time of at least $\lceil \log_r(2n) \rceil$ time units for the calculation of $a+b \pmod{2^n}$. Even with regard to binary digit codings, using a relatively simple addition algorithm (carry look ahead addition), this limit can be closely approximated. We will demonstrate this below.

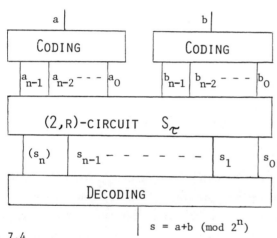

Fig. 7.4

((2,r)-addition circuit)

__Definition 7.4.__ $d_i := a_i \vee b_i; \quad e_i := a_i \oplus b_i; \quad k_i := a_i \cdot b_i \ .$
Let $t(\alpha)$ be the cycle time for calculating α with (2,r)-elements.

With these definitions, we have:

$$s_i = e_i \oplus c_{i-1}, \quad \text{where} \quad c_i := k_i \vee d_i \cdot c_{i-1} \quad (c_{-1} := 0).$$

Since calculation of the sum and carry bit on the far left is the most time-consuming (the remaining binary digits can be calculated at least as quickly in any case), we obtain for the cycle time of the $(2,r)$-circuit required:

$$\tau_n = \max\{t(s_{n-1}), t(s_n)\} \ .$$

Since $s_{n-1} = e_{n-1} \oplus c_{n-2}$ and $s_n = c_{n-1} = k_{n-1} \vee d_{n-1} \cdot c_{n-2}$ we obtain further for $n \geq 2$:

a. $t(s_n) = t(c_{n-1}) \leq \begin{cases} t(c_{n-2}) + 1 & \text{where } r \geq 3 \\ t(c_{n-2}) + 2 & \text{where } r = 2 \ ; \end{cases}$

b. $t(s_{n-1}) \leq t(c_{n-2}) + 1 \ .$

c. We thus obtain for the cycle time of the addition circuit:

$$\tau_n \leq \max\{t(c_{n-1}), t(c_{n-2}) + 1\} \ .$$

Reason. If $r \geq 3$, then c_{n-1} can be calculated from c_{n-2} in one time unit by one $(2,r)$-element from the given formula; if only elements with 2 inputs are available ($r = 2$), 2 time units are necessary for this. Calculation of the auxiliary functions k_i, d_i, e_i is carried out parallel with determination of c_{n-2}, and therefore no extra cycle time is necessary. Since s_{n-1} and s_n may be calculated faster using other formulae, the expressions given for s_n and s_{n-1} are <u>upper</u> time limits.

It is evident that more favourable values can be obtained for the addition time by accelerated calculation of the carries.

Definition 7.5. $L(x) := \begin{cases} 0 & \text{when } x = 0 \\ \lceil \log_r(x) \rceil & \text{when } x \geq 1. \end{cases}$

Lemma 7.7. Let $m = m_1 \cdot g - h$ ($m_1, g \in N$, $0 \leq h \leq g-1$). We have:

$$t(c_{m-1}) \leq 2 + L(m_1) + \max\{t(c_{g-1})-1,\ L(g) + L(m_1-1)\}\ .$$

Proof. We will accelerate calculation of c_{m-1} in accordance with the carry look ahead principle (see 2.3). First, we have:

$$c_{m-1} = k_{m-1}\ v\ d_{m-1}k_{m-2}\ v\ldots v\ d_{m-1}d_{m-2}\ldots d_1 k_0\ .$$

We will use the following auxiliary functions for evaluating this formula:

$$K_{i-1} := k_{ig-1}\ v\ d_{ig-1} \cdot k_{ig-2}\ v\ \ldots\ v\ d_{ig-1} \cdot \ldots \cdot d_{(i-1)g+1} \cdot k_{(i-1)g}$$

$$D_{i-1} := d_{ig-1}\ v\ \ldots\ v\ d_{(i-1)\cdot g} \qquad (i = 1,\ldots,m_1)$$

$$U_i := \begin{cases} D_{m_1-1} \cdot \ldots \cdot D_{i+1} & (i = 0,\ldots,m_1-2) \\ 1 & (i = m_1-1) \end{cases}$$

$$F_i := K_i \cdot U_i \qquad (i = 0,\ldots,m_1-1).$$

If $m = m_1 \cdot g$, then:

$$c_{m-1} = K_{m_1-1}\ v\ D_{m_1-1} \cdot K_{m_1-2}\ v\ D_{m_1-1} \cdot D_{m_1-2} \cdot K_{m_1-3}\ v\ \ldots$$

$$v\ \ldots\ v\ D_{m_1-1} \cdot D_{m_1-2} \cdot \ldots \cdot D_1 \cdot K_0$$

$$= U_{m_1-1} \cdot K_{m_1-1}\ v\ U_{m_1-2} \cdot K_{m_1-2}\ v\ U_{m_1-3} \cdot K_{m_1-3}\ v\ \ldots\ v\ U_0 \cdot K_0$$

$$= F_{m_1-1}\ v\ \ldots\ v\ F_0\ .$$

If $m = m_1 \cdot g - h$ ($0 < h \leq g-1$), the definition of the auxiliary values K_{m_1-1} and D_{m_1-1} must be changed. This will not result in a longer cycle time as compared with the calculation of c_{m-1} ($m = m_1 \cdot g$).

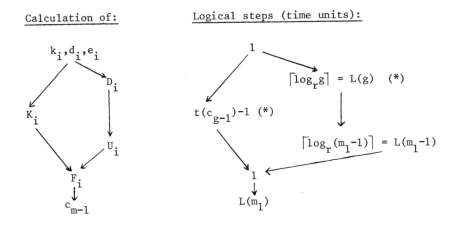

Fig. 7.5

Re.(*): The time required for K_i is as great as for $t(c_{g-1})$ for calculating the total carry of two g-digit numbers minus one logical step for determination of k_i, d_i and e_i, which has been carried out already. For constructing an AND or OR gate with p inputs from (d,r)-elements we need $\lceil \log_r p \rceil = L(p)$ logical steps.

Since calculation of K_i can be carried out in parallel with determination of D_i and U_i, fig. 7.5 yields:

$$t(c_{m-1}) < 1 + \max\{t(c_{g-1})-1, \; L(g)+L(m_1-1)\} + 1 + L(m_1).$$

We can now indicate a limit for the maximum cycle time for calculating one carry:

Lemma 7.8. *If* $v_k := r^{k(k-1)/2}$ $(k \in \mathbb{N})$, *then:*

There is a (2,r)-circuit with:

$$t(c_{v_k-1}) \leq 1 + \frac{k(k+1)}{2} .$$

Proof. We will demonstrate this by induction according to k. The proposition applies to $k = 1$ (i.e. $v_1 = r^0 = 1$), since:

$$t(c_{1-1}) = t(c_0) = t(e_0 \vee d_0 \cdot c_{-1}) \leq 2 .$$

(It is evident that the proposition remains correct even for subtraction, using the 2-complement $(c_{-1} = 1)$).

If $t(c_{v_k-1}) \leq 1 + \frac{k(k+1)}{2}$, and since

$$v_{k+1} = r^{k(k+1)/2} = v_k \cdot r^k$$

by application of lemma 7.7 with $m_1 = r^k$ and $g = v_k$ we obtain the relationship:

$$t(c_{v_{k+1}-1}) \leq 2 + L(r^k) + \max\{t(c_{v_k-1})-1, \ L(v_k)+L(r^k-1)\}$$

$$\leq 2 + L(r^k) + \max\{\frac{k(k+1)}{2}, \ L(v_k)+L(r^k-1)\}$$

$$= 2 + k + \max\{\frac{k(k+1)}{2}, \ \frac{k(k-1)}{2} + k\}$$

$$= 2 + k + \frac{k(k+1)}{2} = 1 + \frac{(k+1)(k+2)}{2} .$$

Q.e.d.

If $n \geq 3$, there is a uniquely defined $k \in \mathbb{N}$ with

$$r^{(k-1)(k-2)/2} < n-1 \leq r^{k(k-1)/2}$$

i.e.
$$\frac{(k-1)(k-2)}{2} < \log_r(n-1) \leq \frac{k(k-1)}{2} \ .$$

For this uniquely defined value k which is dependent of n, we obtain:

$$\max\{t(c_{n-2}), \ t(c_{n-1})\} \leq 1 + \frac{k(k+1)}{2} \ .$$

Hence, for the total cycle time τ_n of the addition circuit, it follows that:

$$\tau_n \leq \max\{t(c_{n-2})+1, \ t(c_{n-1})\} \leq 1 + 1 + \frac{k(k+1)}{2}$$

$$= \frac{(k-1)\cdot(k-2)}{2} + 2k+1 \approx \log_r(n-1)+2k+1 \ .$$

Applying the upper limit for k finally yields:

$$\tau_n < \log_r(n-1) + 2(\sqrt{2\log_r(n-1) + \frac{1}{4}} + \frac{3}{2}) + 1$$

$$= (1 + \varepsilon_n)\cdot\log_r(n-1) \qquad (\varepsilon_n \xrightarrow[n\to\infty]{} 0) \ .$$

This provides the following result [Br1] which should be compared to the corresponding result of 7.2.2:

<u>Proposition 7.9.</u> *For each $r \geq 2$ and each $\varepsilon > 0$ there is a $n_0 = n_0(\varepsilon, r)$ with:*

For each $n \geq n_0$ there is a $(2,r)$-circuit for the addition of two n-digit numbers, for which the addends and results exist in binary digit coding, and whose cycle time is:

$$\tau_n < (1+\varepsilon)\cdot \log_r(n-1) < (1+\varepsilon)\cdot \lceil \log_r 2n \rceil \ .$$

The cycle time of this circuit - based on the carry look ahead addition principle - is therefore greater only by less than a factor $(1+\varepsilon)$ than the generally valid lower estimate of cycle time, as derived under proposition 7.5 and in 7.2.2. (example A).

BIBLIOGRAPHY

[An1] Anderson, D.W.
       Sparacio, F.J.
       Tomasulo, R.M.
       The IBM System/360 Model 91: Machine Philosophy and Instruction-Handling.
       IBM Journal Res. Dev. $\underline{11}$ (1967) 8-24

[An2] Anderson, S.F.
       Earle, J.G.
       Goldschmidt, R.E.
       Powers, D.M.
       The IBM System/360 Model 91: Floating-Point Execution Unit.
       IBM Journal Res. Dev. $\underline{11}$ (1967) 34-53

[At1] Atkins, D.E.
       Higher Radix Division Using Estimates of the Divisor and Partial Remainder.
       IEEE-C $\underline{17}$ (1968) 925-934

[At2] Atkins, D.E.
       Design of the Arithmetic Units of ILLIAC III. Use of Redundancy and Higher Radix Methods.
       IEEE-C $\underline{19}$ (1968) 720-732

[Av1] Avizienis, A.
       Signed-Digit Number Representations for Fast Parallel Arithmetic.
       IRE-EC $\underline{10}$ (1961) 389-400

[Av2] Avizienis, A.
       On a Flexible Implementation of Digital Computer Arithmetic.
       IFIP (1962) 664-670

[Av3] Avizienis, A.
       Tung, C.
       A Universal Arithmetic Building Element and Design Methods for Arithmetic Processors.
       IEEE-C $\underline{19}$ (1970) 733-748

[Ba1] Banerji, D.K.
       Brzozowski, J.A.
       On Translation Algorithms in Residue Number Systems.
       IEEE-C $\underline{21}$ (1972) 1281-1285

[Ba2] Banerji, D.K.
       On the Use of Residue Arithmetic for Computation.
       IEEE-C $\underline{23}$ (1974) 1315-1317

[Be1] Bedrij, O.J.
       Carry-Select-Adder.
       IRE-EC $\underline{11}$ (1962) 340-346

[Br1] Brent, R.
       On the Addition of Binary Numbers.
       IEEE-C $\underline{19}$ (1970) 758-759

[Bu1]	Buchholz, W.	Planning a Computer System. McGraw-Hill (1962)
[Ca1]	Cappa, M. Hamacher, V.C.	An Augmented Iterative Array for High-Speed Binary-Division. IEEE-C 22 (1973) 172-175
[Ch1]	Chen, T.C.	Automatic Computation of Exponentials, Logarithms, Ratios and Square Roots. IBM Journal Res. Dev. 16 (1972) 380-388
[Cl1]	Claus, V.	The Average Period of Addition of a Parallel Adder (Die mittlere Additions- dauer eines Paralleladdierwerkes). Acta Informatica 2 (1973) 283-291
[Da1]	Dadda, L.	Some schemes for parallel multipliers. Alta Frequenza 34 (1965) 349-356
[Da2]	Dadda, L. Ferrari, D.	Digital Multipliers: A Unified Approach Alta Frequenza 37 (1968) 1079-1086
[De1]	Dean, K.J.	Design for a full multiplier. Proc. IEEE 115 (1968) 1592-1594
[Ea1]	Earle, J.	Latched Carry-Save-Adder. IBM Tech. Dis. Bull. 7 (1965) 909-910
[Fe1]	Ferrari, D.	A Division Method Using a Parallel Multiplier. IEEE-EC 16 (1967) 224-226
[Fe2]	Ferrari, D.	Fast Carry-Propagation Iterative Networks. IEEE-C 17 (1968) 136-145
[Fe3]	Ferrari, D. Stefanelli, R.	Some new schemes for parallel multipliers. Alta Frequenza 37 (1969) 843-852
[Fl1]	Flores, I.	The Logic of Computer Arithmetic. Prentice-Hall (1963)

[Fl2]	Flynn, M.J.	Very High Speed Computing Systems. Proc. IEEE 54 (1966) 1901-1909
[Fl3]	Flynn, M.J.	On Division by Functional Iteration. IEEE-C 19 (1970) 702-706
[Fr1]	Freiman, C.V.	Statistical Analysis of Certain Binary Division Algorithms. Proc. IRE 49 (1961) 91-103
[Ga1]	Garner, H.L.	The Residue Number System. IRE-EC 8 (1959) 140-147
[Ga2]	Garner, H.L.	Number Systems and Arithmetic. Adv. in Comp. 6 (1965) 131-194
[Go1]	Gosling, J.B.	Design of large high-speed floating-point-arithmetic units. Proc. IEE 118 (1971) 493-498
[Go2]	Gosling, J.B.	Design of large high-speed binary multiplier units. Proc. IEE 118 (1971) 499-505
[Ha1]	Habibi, A. Wirtz, P.A.	Fast Multipliers. IEEE-C 19 (1970) 153-157
[Ha2]	Hallin, T.G. Flynn, M.J.	Pipelining of Arithmetic Functions. IEEE-C 21 (1972) 880-886
[He1]	Hendrickson, H.C.	Fast High-Accuracy Binary Parallel Addition. IRE-EC 9 (1960) 465-469
[Ho1]	Hotz, G.	Informatik. Rechenanlagen. Teubner Studienbücher (1972).
[Hu1]	Husson, S.S.	Microprogramming: Principles and Practices. Prentice Hall (1970)
[Ka1]	Kamal, A.A. Ghannam, M.A.N.	High-Speed Multiplication Systems. IEEE-C 21 (1972) 1017-1021

[Ki1]	Kilburn, T. Edwards, D.B.G. Aspinall, D.	Parallel Addition in Digital Computers; A New Fast Carry Circuit. Proc. IEE __106 B__ (1959) 464-466
[Ki2]	Kilburn, T. Edwards, D.B.G. Aspinall, D.	A Parallel Arithmetic Unit using a Saturated-Transistor Fast-Carry Circuit. Proc. IEE __107 B__ (1960) 573-584
[Ki3]	Kinniment, D.J. Steven, G.B.	Sequential-state binary parallel adder. Proc. IEE __117__ (1970) 1211-1218
[Kl1]	Klar, R.	Digitale Rechenautomaten. De Gruyter (1976)
[Kr1]	Krishnamurthy, E.V.	On Optimal Iterative Schemes for High- Speed Division. IEEE-C __19__ (1970) 227-231
[Le1]	Lehman, M. Burla, N.	Skip Techniques for High-Speed Carry- Propagation in Binary Arithmetic Units. IRE-EC __10__ (1961) 691-698
[Le2]	Lehman, M.	A Comparative Study of Propagation Speed-Up Circuits in Binary Arithmetic Units. IFIP (1962) 672-676
[Le3]	Lewin, D.	Theory and Design of Digital Computers. Nelson (1972)
[Li1]	Ling, H.	High-Speed Computer Multiplication Using a Multiple-Bit Decoding Algorithm. IEEE-C __19__ (1970) 706-709
[Ma1]	MacSorley, O.L.	High-Speed Arithmetic in Binary Computers. Proc. IRE __49__ (1961) 67-91
[Ma2]	Majerski, S.	On Determination of Optimal Distributions of Carry Skip in Adders. IEEE-EC __16__ (1967) 45-58
[Me1]	Meggitt, J.E.	Pseudo Division and Pseudo Multiplication Processes. IBM Journal Res. Dev. __6__ (1962) 210-226

[Me2] Meo, A.R.   Arithmetic Networks and Their Minimization
                                        Using a New Line of Elementary Units.
                                        IEEE-C 24 (1975) 258-280

[Pr1] Pradhan, D.K.   Fault-Tolerant Carry-Save-Adders.
                                        IEEE-C 23 (1974) 1320-1322

[Ra1] Ramamoorthy, C.V.   Fast Multiplication cellular arrays
       Economides, S.C.   for LSI implementation.
                                        Fall Joint Comp. Conf. (1969) 89-98

[Ra2] Ramamoorthy, C.V.   Some Properties of Iterative Square-
       Goodman, J.R.   Rooting Methods Using High-Speed Multi-
       Kim, K.H.   plication.
                                        IEEE-C 21 (1972) 837-847

[Ro1] Robertson, J.E.   A New Class of Digital Division Methods.
                                        IRE-EC 7 (1958) 218-222

[Ro2] Robertson, J.E.   The Correspondence Between Methods of
                                        Digital Division and Multiplier Recording
                                        Procedures.
                                        IEEE-C 19 (1970) 692-701

[Sa1] Salter, F.   High-speed transistorized Adder for a
                                        Digital Computer.
                                        IRE-EC 9 (1960) 461-664

[Sa2] Sarkar, B.P.   Economic Pseudodivision Processes for
       Krishnamurthy, E.V.   Obtaining Square Roots, Logarithms and
                                        Arc tan.
                                        IEEE-C 20 (1971) 1589-1593

[Sa3] Sasaki, A.   Addition and Subtraction in the Residue
                                        Number System.
                                        IEEE-EC 16 (1967) 157-164

[Sk1] Sklansky, J.   Conditional-Sum Addition Logic.
                                        IRE-EC 9 (1960) 226-231

[Sp1] Spira, P.M.   The Time Required for Group Multiplication.
                                        Jour. Ass. Comp. Mach. 16 (1969) 235-243

[St1] Stefanelli, R.   A Suggestion for a High-Speed Parallel Binary Divider.
IEEE-C 21 (1972) 42-55

[Sw1] Swartzlander, E.E.   The Quasi-Serial Multiplier.
IEEE-C 22 (1973) 317-321

[To1] Tocher, K.D.   Techniques of Multiplication and Division for Automatic Binary Computers.
Quart. J. Mech. Appl. Math. 11 (1958) 364-348

[To2] Tomasulo, R.M.   An Efficient Algorithm for Exploiting Multiple Arithmetic Units.
IBM Journal Res. Dev. 11 (1967) 25-33

[Tu1] Tung, C.   A Division Algorithm for Signed-Digit Arithmetic.
IEEE-EC 17 (1968) 887-889

[Tu2] Tung, C.   Arithmetic.
In: Cardenas, A.F.   Computer Science.
     Presser, L.   Wiley (1972) 59-102
     Marin, M.A.

[Vo1] Volder, J.E.   The CORDIC Trigonometric Computing Technique.
IRE-EC 8 (1959) 330-334

[vN1] v. Neumann, J.   Preliminary Discussion of the Logical Design of an Electronic Computing Instrument.
      Burks, A.W.
      Goldstine, H.H.   Collected Works 5 (1961) 34-80

[Wa1] Wallace, C.S.   A Suggestion for a Fast Multiplier.
IEEE-EC 13 (1964) 14-17

[Wa2] Walther, J.S.   A Unified Algorithm for Elementary Functions.
AFIPS SJCC 38 (1971) 379-385

[Wi1] Wilkinson, J.H.   Rounding Errors in Algebraic Processes.
Springer Verlag (1969).

[Wi2] Winograd, S. On the Time Required to Perform Addition.
J. Assoc. Comp. Mach. 12 (1965) 277-285

[Wi3] Winograd, S. On the Time Required to Perform Multiplication.
J. Assoc. Comp. Mach. 14 (1967) 793-802

INDEX

Adder-tree 44,46
addition 15,16
- with representation by A+S 17
-, carry look ahead 49,261
-, CLA, 1st order 51
-, CLA, 2nd order 54
-, carry ripple 45
-, --, asynchronous 46
-, --, synchronous 45
-, carry save 41
-, carry select 72
-, carry skip 56
-, conditional sum 67
-, d-complement 14
-, (d-1)-complement 16
-, exclusive-OR 47
-, serial 38
-, von Neumann 39
algorithm
-, Booth 85
arc tan 223
arithmetic circuit 120
A+S 11

Bailey iteration 176
basic choice 20
binary SRT division 211
Booth algorithm 85

Calculation 254
carry completion 46
carry-in 41
carry look ahead addition 49
-, 1st order 51
-, n-bit 49
-, 2nd order 54
carry
- ripple addition 45
---, asynchronous 46
---, synchronous 45
- save addition 41
- select addition 72
- skip addition 56
- propagation 130
-, total 14,51,54
CSA steps 42
characteristic sequence (CS) 122
circuit 21
-, arithmetic 120
-, compact 122,127
-, triangular 122, 125
coding 246
compact circuit 122,127
connection 21
convergence 174
-, direction of 180
-, speed of 233
conversion 196,199
CORDIC method 228,229,233,238

costs   20,33,75
counter
-, (3,2) -   35,108
-, (d,r) -   260
-, (m,k) -   33
-, (4.2) -   35,119
-, (2.2) -   34,108
cycle time   33,75

Decoding   245
deconversion   19,198
denary digit coding   10
digit coding   10
-, denary   10
digit system   14
dirty zero   28
disjunctive normal form   35
dividend register   140
division
-, Anderson   179
-, Ferrari   179
-, iterative   173
-, Newton   175
-, non-performing   144
-, non-restoring   144
-, pseudo   218,219,221,223,225
-, restoring   142
-, SRT   199,209
--, binary   211
-, table look up   155,173,180
divisor
- register   140
- multiples   157

double precision   24
doubling of the sign   13

Earle latch   134,189
element, <-1,0,3,2> -   127
end around carry   37
errors   174
evolution method   191,224
exclusive-OR addition   47
exponent   25
-, overflow   26
-, underflow   26

Feedback   136
fixed-point representation   22,79
floating-point representation   25
formal sum   14
functions
-, hyperbolic   238
-, inverse   226
-, trigonometric   230
fulladder   33,35,37,102,104

Generalized Newton method   175,177
group size   51
group scheme   60
group carry   53,71

Halley iteration   176
halfadder   33,34,102,104
h_j-separable   248
hyperbolic functions   238

Input sequence 122
insertion 12
inverse function
-, calculation of 226
iteration
-, Bailey 176
-, Halley 176
iteration errors 174
iteration methods 174
iteration rules 174
iterative division 173

Latch 134,189
-, Earle 134,189
limit value 174
logarithm
-, calculation of 218
loop 21

Mantissa 25
Meggitt method 219
method
-, Anderson 179
-, Briggs 218
microprogram 20
minimal redundant 195
(m,k)-counter 33
modular design 33
multiplicand register 77

multiplication
-, Dadda 109
-, Wallace 106
-, pseudo 218,222,225
multiplication matrix 100
multiplier
- coding 88,92
- register 77
multiplication method
-, serial 80
-, parallel 100

n-bit carry look ahead addition 49
Newton method 175,177
-, division free 192
-, generalized 175,177
non-normalized floating-point
    representation 25
non-redundant 11
non-performing division 144
non-restoring division 144
normalisation 25
-, pre- 26
-, post- 27
notation 10
-, redundant 19,195

Optimum base 20
order
-, carry look ahead, 1st 51
-, carry look ahead, 2nd 54
- of the iteration method 174

overflow   13
- recognition   13,15,17,202
- correction   202
- criterion   13
- problem   12,78
output sequence   122

Partial
- product   80
--, reduced   80
- remainder   221
pipelining   131,173,187
postnormalisation   27
postshift   27
prenormalisation   26
preshift   25
propagation chain   46

Quotient register   140

Reduction of M_0 102
redundant notation   19,195
redundant
-, maximal   195
-, minimal   195
register   20
- configuration   140
- transfer   21

representation
- by amount and sign   11
-, d-complement   11
-, (d-1)-complement   12
-, fixed-point   22,79
-, floating-point   25
--, non-normalized   25
-, SDNR   195
residue arithmetic   19
restoring division   142
roots   224,238
r-reduced   125
rounding   23,79,143
rounding error   22

SDNR   195
serial addition   38
section   54
shift
- over ones   87,148
- over zeros   87,147
- size   85
--, mean   158,161,167
- tables   159,160,166
scaling   22,80,140
special functions   218
square roots   191
SRT division   199,209
-, binary   211
starting value   175,180
subtraction   37

system
-, (1/2,1,2) - 158
-, (3/4,1,3/2) - 165
-, (3/4,1,5/4) - 166
-, (5/8,1,5/4) - 166

Table look up division   155,173,186
Taylor development   174
total carry   14,51,54
transformation rule   229
triangular circuit   122,125
trigonometric functions   230

von Neumann addition   39

Wallace
- -tree   44,106,136
- multiplication   106
weight
-, absolute   123
word length   26,31,32

SYMBOLS

N	Set of natural numbers
N_0	$N \cup \{0\}$
Z	Set of integers
Q	Set of rational numbers
R	Set of real numbers
B_d	$\{0,\ldots,d-1\}$
$A \times B$	Cartesian product
$|a|$	Value of a
$\lfloor x \rfloor$	$\max\{n \mid n \in Z;\ x \geq n\}$
$\lceil x \rceil$	$\min\{n \mid n \in Z;\ x \leq n\}$
$O(n)$	Landau symbol
$\circ$	Sequential operation
$\dot{x}$	Parallel operation } see 3.4
DD	Dividend register
DR	Divisor register
MD	Multiplicand register
MQ	Multiplier register
MP	Partial product register
FA	Fulladder
HA	Halfadder
τ, t	Cycle time
κ	Cost function
$P^{(j)}$	Partial product
$Q^{(j)}$	Reduced partial product
$X^{(j)}$	Partial remainder
q_j	Quotient bit
$\text{sign}(x)$	Sign of x
$x, \cdot$	Multiplication sign
$a \wedge b$	Logical AND $(a \wedge b = 1 \Leftrightarrow a = b = 1)$
$a \vee b$	Logical OR $(a \vee b = 0 \Leftrightarrow a = b = 0)$
$\bar{a}$	Negation $(\bar{a} = 1 \Leftrightarrow a = 0)$
$a \oplus b$	a+b mod 2 (logical E x OR)

Q	10	$\hat{\oplus}$	52
$w^{(n,m)}$	10	$\hat{\hat{\oplus}}$	54
$B_d = \{0,1,\ldots,d-1\}$	10	MD	77
A+S	11	MQ	77
w_d	11	MP	77
w_{A+S}	11	MH	77
w_{d-1}	12	$P^{(j)}$	80
$Q_A^{(n,m)}$	13	$Q^{(j)}$	80
C	20	SHR(A)	81
$MP^{(1)}, MP^{(2)}$	22	MQ_{-1}	86
d_a	29	$n_{(3,2)}$	104
d_e	29	$\kappa_{(3,2)}$	104
s	33	$n_{(2,2)}$	104
c	33	$\kappa_{(2,2)}$	104
$f_{(3,2)}$	33	n_{OR}	104
$f_{(2,2)}$	33	κ_{OR}	104
$\oplus$	33	CS	122
τ_{SK}	34	DD	140
κ_{SK}	34	DR	140
HA	34	DE	140
FA	35	(DD,DE)	140
CSA	42	$X^{(j)}$	141

q_j	141
SHL(A)	143
$\varphi(x_i)$	174
$(2-d_i)_{T_r}$	183
$w_{red}^{(n,m)}$	194
λ_i	230
δ_i	230
α_i	230
Φ	245
h'	245
S_τ	245
h	247